SpringerBriefs in Latin American Studies

More information about this series at http://www.springer.com/series/14332

Nathan Einbinder

Dams, Displacement and Development

Perspectives from Río Negro, Guatemala

CLAG
clagscholar.org

Springer

Nathan Einbinder
Department of Agroecology and Society
Colegio de la Frontera Sur
San Cristóbal de las Casas
Chiapas
Mexico

ISSN 2366-763X ISSN 2366-7648 (electronic)
SpringerBriefs in Latin American Studies
ISBN 978-3-319-51510-6 ISBN 978-3-319-51511-3 (eBook)
DOI 10.1007/978-3-319-51511-3

Library of Congress Control Number: 2017932119

Printed on acid-free paper

This Springer imprint is published by Springer Nature
The registered company is Springer International Publishing AG
The registered company address is: Gewerbestrasse 11, 6330 Cham, Switzerland

Preface

To Witness the Rage and Silence of Genocide in Río Negro and Pacux
By Catherine Nolin

> "Bearing witness," like "solidarity" and "compassion" is a term worth rehabilitating.... Bearing witness is done on behalf of others, for their sake (even if those others are dead and forgotten).... No matter how great the pain of bearing witness, it will never be as great as the pain of those who endure, whether in silence or with cries...
>
> Paul Farmer, Bearing Witness (2005, 28)

To Witness

To witness, listen, and write well about the cries and silences of life in contemporary Río Negro and Pacux is to write about post-genocide realities of struggle and strength. It is to write about pain and social suffering, on the one hand, and brutality and structural violence on the other. To carry out this daunting task, Nathan Einbinder opened his mind to critical questions of the deep connections between 'development' and violence and grappled with ensuing violent geographies at the scale of the individual to those at the scale of the community, nation, and onto the dominant transnational worlds of the World Bank, Inter-American Development Bank, and the American Government.

Nathan also opened his heart to the *testimonios* of community members and survivors, those who rage against the decades of injustice that followed displacement and genocide, as well as those often silenced in post-conflict realities of trauma, hunger, and dysfunction. In the pages to follow, Nathan (*final page, final paragraph) demonstrates what it looks like to bear witness as an academic and activist committed to 'proclaiming that it is our duty to alert the international community of what is happening in the far corners in which we study.'

To Witness Together

Grahame Russell of Rights Action and I have co-organized delegation-style field schools to Guatemala since 2004 as part of our vision of transformative learning. As a former graduate student of mine, Nathan traveled with fellow university students and us many times over the years to explore issues of culture, rights, and power through the experiences of state violence, genocide, and recovery in Guatemala. The most honorable way to do this work is through a commitment to the very human action of witnessing together. Witnessing is more than observing. *Témoignage*—the French word for witnessing—is the act of being willing to speak out about what we see happening in front of us. Bearing witness in the context of human rights atrocities and major crimes means a willingness to speak on behalf of the people we meet, to amplify their voices, and to bring abuses and intolerable situations to the public eye. This book is Nathan's witness.

To Witness Together in Río Negro and Pacux

Back in 2008, two years before Nathan started this journey, Sebastian Iboy Osorio, President of the COCODE (community development organization) of Río Negro, quietly arrived in our hotel in Rabinal with his son Nelson while we listened to a talk by Fernando Suazo, a former activist Spanish priest who arrived here in 1984 (shortly after the series of massacres that hit Rabinal like a tidal wave in 1982). Fernando left the church and married into the Maya Achí culture and so offers a window into the Maya cosmovision in a way that few non-Maya can. We talked liberation theology, big business, neocolonization, and the process of recovering the historical memory of the years of genocide, silence, the culture of violence, and so much more.

Sebastian—I'm not sure how to put into words what a beautiful person he is. Sebastian was only 16 years old when, in 1982, his mother, father, sisters, brothers, cousins, aunts, uncles, and friends were ripped out of his life in a whirlwind of brutal violence. First, his father and brother were murdered in the Xococ massacre of February 13, 1982, not far from his home community (he wanted to go with them as they were called by the patrols to present themselves in Xococ to be accounted for by the military commissioner. But, his father told him that there might be violence, they may even be killed, so to please stay at home with his family). A month later, he was an eyewitness to the murder of 177 women and children on March 13, 1982, up the mountain from his home in Río Negro.

Sebastian recounted these events and the years of hiding in the mountains with Carlos Chen and others, meeting and marrying his wife Magdalena in a ditch while in hiding, and his eventual return to Río Negro a few years before to help with the REMHI project (the Catholic Church-driven recovery of historical memory project to document the human rights violations and genocide). On Mother's Day, we awoke at 5:30 am to make the long, difficult, physically challenging walk up to that massacre site with Sebastian. The significance of making this journey with Sebastian on Mother's Day to the place where so many women and children were brutally ripped from this world was not lost on our group. We were hot, tired, needing water, needing rests … but always knowing that the women and children were forced to march up this mountain at gunpoint with no rest, no water, no hope. Sebastian quietly showed us where he hid, where he saw this horror took place, what he and Carlos found when they crept back a day later. He has given his *testimonio* before so he could recount these events calmly and clearly. But, of course, we could hardly breathe, we could barely speak.

Sebastian looked straight into the eyes of each of us and said that he thanked us a thousand times for being with him that morning. That he lost so much in this life and that our presence validated that it was not right, what happened was wicked, and that our connection to the place would be forever.

Sebastian, Cornelio, Cristobal and so many others came to put their trust in Nathan in the following years to bear witness to the story of Río Negro and Pacux. He is now part of the remembering, the resisting, and the hope for something better.

To Witness and Accompany

Through his devoted accompaniment of the communities of Río Negro and Pacux, Nathan is clear that the well-documented human rights violations cited here are 'not accidents; they are not random in distribution or effect. Rights violations are, rather, a symptom of deeper pathologies of power and are linked intimately to the social conditions that so often determine who will suffer abuse and who will be shielded from harm' (Farmer 2005, xiii). These social/political/economic/moral conditions and their devastating effects sit at the heart of Nathan's critique of power, racism, extreme inequality, and neoliberal visions of 'development' in Guatemala. And they also offer the grounds on which truth- and justice-seeking activism emerges, grows, is threatened, moves forward.

The unequal power we witness in Guatemala—with a set of military and elite alliances that, by all measures, have ferociously targeted, repeatedly, a largely defenseless Indigenous and progressive Ladino majority population—generates brutality that is both obvious but quiet on the world stage. Through documentary analysis and face-to-face conversations with survivors in Río Negro and Pacux, Nathan reveals some of the logic of power and the expected outcomes of state-sponsored terror and total disregard for the human dignity of the indigenous Maya population. Structural violence, internal to the country and supported beyond the country's borders, has, for centuries, enabled an intransigent elite/military/political minority to benefit from the suffering of the majority, and even call it 'development.'

Throughout this book, Nathan Einbinder illustrates the devastating impacts of structural violence, of the still-shocking collaborations among the Guatemalan state/military/elite, the World Bank, American decision-makers and funders, among others. Their roles as the intellectual and material authors of the Chixoy Dam 'development' project and the crimes linked to it continue to go unpunished but change is happening.

We are all witnesses now, through the *testimonios* that Nathan gathered, to courageous struggles for dignity, truth, and justice.

Dr. Catherine Nolin
Department of Geography
University of Northern British Columbia
Prince George, British Columbia
Canada

Acknowledgements

First and foremost, I would like to thank my supervisor for this work, Dr. Catherine Nolin. Without her support, mentorship, and encouragement, none of this would have been possible. I feel honored for the opportunity to share two and a half years with Catherine and her crew of graduate students, and to witness her dedication to issues of injustice in Guatemala and Canada. She is an inspiration to us all.

Secondly, I thank the survivors of the Río Negro massacres and their families, for sharing their history and a small piece of their lives. I sleep better at night knowing their persistence and strong belief, and I hope to remain connected to them, and their cause for justice, for as long as I can. Along with community members in Río Negro and Pacux, I thank Heidi McKinnon of the Advocacy Project, who facilitated many of my interviews and meetings. Without her deep connection to the women of Pacux, I doubt that my interviews would have been as productive and moving.

Words of gratitude and respect for Grahame Russell of Rights Action could fill its own page. Without his support, humor, introductions, and knowledge of the issues, I am not sure where I would be with all of this. It is through his years of work and dedication to the people of Río Negro that made this project possible.

There are many others who made this project and its completion possible. I thank my committee members Matthew Taylor and Zoe Meletis for their support, encouragement, time, and patience. I would also like to thank the entire 'Nolin group,' particularly J.P. Laplante and Claudette Bois, who both took far too much time to help me with drafts and formulating ideas.

Finally, I thank the UNBC Office of Research and Graduate Studies, who provided funding for my research in 2009 and for my follow-up visit to Río Negro in 2010.

Contents

Timeline of Events at Río Negro

June 1972—Guatemala's National Institute of Electrification (INDE), with the help of private Consorcio LAMI, a German-based construction firm, identifies the Chixoy Basin as the site for a new hydroelectric dam. Feasibility studies begin.

January 1976—Inter-American Development Bank (IDB) loan US$105 million of granted to INDE for the construction of Chixoy.

1976–78—Army and INDE begin to show up at Río Negro and other Chixoy-affected communities to notify them of their displacement.

June 1978—World Bank loan of US$72 million granted to INDE for Chixoy.

January 1980—Leaders from Río Negro are taken to Pacux, the resettlement community begin constructed for their community near Rabinal. After viewing the poor construction and lack of arable land, they reject the compensation plan offered by INDE.

March 1980—Tensions rise between military police and Río Negro residents after two workers (both from Río Negro) are arrested and tortured after being accused of stealing beans. When police arrive to the village for further interrogation, a skirmish takes place leaving seven Río Negro residents dead from gunfire. One of the military police officers also died from drowning in the river after being attacked by angry citizens. From this time on, the army arrives repeatedly to the village, interrogating and torturing residents in attempt to get information about guerrillas.

Spring 1980—INDE demands Río Negro citizens to hand over land titles, promising to return them. Later in the year INDE claims to have never received them.

July 1980—INDE demands the Libro de Actos to be delivered to the dam site by village leaders. This book outlines all compensation agreements and promises made by the INDE. The two men delivering the book were kidnapped and murdered. The book was never seen again.

September 1981—Army massacre at an annual fair in Rabinal claim some 200+ innocent civilians.

October 1981—Army massacre in nearby village of Xococ. A civilian defense patrol (PAC) is established.

November 1981—Río Negro forced to form a PAC.

February 4, 1982—Xococ attacked by guerrillas, killing five PAC members. A captured guerrilla names individuals form nearby communities, including Río Negro.

February 6–7, 1982—80 Río Negro residents are summoned to Xococ by the local military commander. They are captured but later freed, after handing over ID cards.

February 13, 1982—Río Negro residents return to Xococ. They are accused of being guerrillas, and subsequently fired upon. 71 individuals are killed. One survivor returns to Río Negro with the news. Army and Xococ PAC members later surround the village, chasing individuals into the surrounding mountains and into other villages.

March 13, 1982—Army and Xococ PAC arrive to Río Negro and round up women and children (the men were in hiding at this time—thinking the women were safe). They are forced up the mountainside to a pass called Pocoxom. 107 children and 70 women are murdered and tossed into a ditch. 18 children are kept as slaves by PAC members. The village is deserted, with survivors either taking shelter in surrounding villages, or living in the mountains.

May 14, 1982—92 Río Negro residents are killed by Army soldiers at the nearby village of Los Encuentros. 15 women are taken away by helicopter, never to be seen again.

June 1982—General Efraín Ríos Montt becomes president resulting from a coup. He offers 'amnesty' to refugees living in the mountains and drops pamphlets from helicopters.

September 14, 1982—Massacre at nearby village of Agua Fría takes the lives of 35 refugee children from Río Negro.

January 1983—Dam gates closed and reservoir begins to fill. August 1983—The first families begin to arrive in Pacux. Many are tortured and kept in the military garrison built adjacent to the community.

March 1985—Another World Bank of US$44 million is granted to INDE, partially for rebuilding damaged tunnels.

April 1986—Commercial use of the facility begins. 1991—Río Negro massacre survivors and their families begin to return the land directly above the old village site of Río Negro, now underwater.

Map

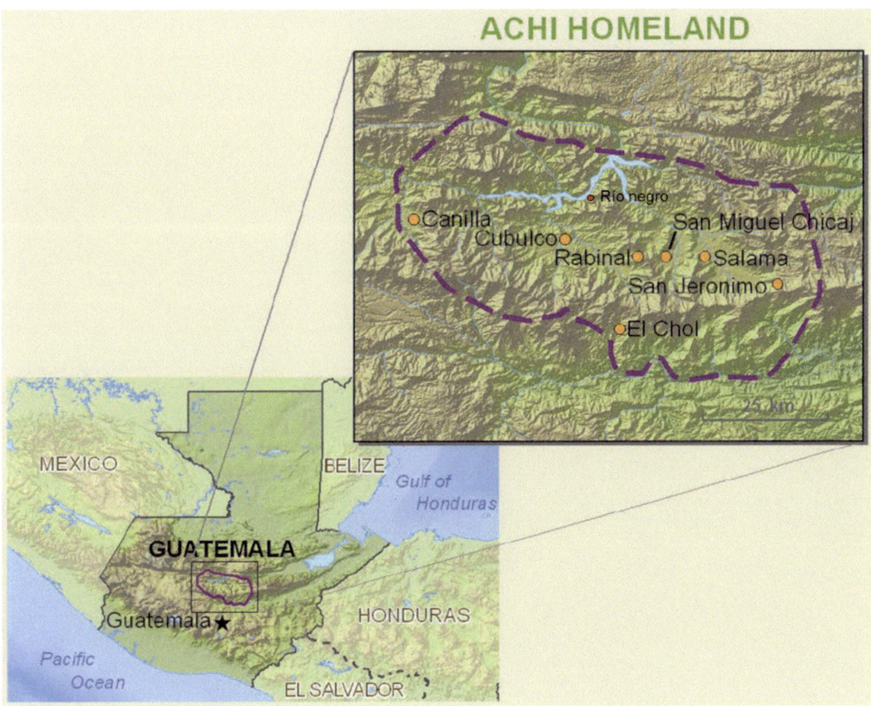
ACHI HOMELAND
Río negro
Canilla
Cubulco
Rabinal
San Miguel Chicaj
Salama
San Jeronimo
El Chol
25 km
MEXICO
BELIZE
Gulf of Honduras
GUATEMALA
Guatemala
HONDURAS
Pacific Ocean
EL SALVADOR

Acronyms

ADIVIMA	Association for the Integral Development of the Victims of Violence in the Verapaces, Maya Achí
ASCARA	Campesino Association Río Negro 13 of March Maya Achí
CAFTA-DR	Central American Free Trade Agreement—Dominican Republic
CEH	Commission for Historical Clarification
COCAHICH	Coordinator of the Communities Affected by the Construction of the Chixoy Dam
COCODE	Community Development Council
CONIC	National Indigenous and Peasant Coalition
COPAE	Guatemalan Pastoral Commission on Peace and Ecology
EGP	Guerrilla Army of the Poor
FAR	Rebel Armed Forces
FRG	Guatemalan Republican Front
GDP	Gross Domestic Product
IDB	Inter-American Development Bank
IMF	International Monetary Fund
INDE	National Institute of Electrification
NAFTA	North American Free Trade Agreement
NGO	Non-governmental Organization
ORPA	Revolutionary Organization of the People in Arms
PAC	Civilian Defense Patrol
UFCo	United Fruit Company
UNDP	United Nations Development Program
URNG	United Revolutionary Unity of Guatemala
USAID	United States Agency for International Development

Chapter 1
Introduction

Abstract The decision to embark on a research project in a place such as Río Negro did not come at random. In this introductory chapter, I explain the personal events that lead to my first trip to the community, as well as the decision to make their history of displacement and violence the basis of my master's thesis. In addition, I present the general approach and layout of this book, as well as a brief summary of the Río Negro story, which I will expand upon at length in the following chapters.

Keywords Latin American history · North–South relations · Critical human geography · Ethnography

My early perceptions and imaginations of Latin America were formed in the border city of San Diego, where I grew up. Aside from being one of the wealthiest regions in the country, if not the world, nowhere, as far as I know, are the contrasts any sharper. Not more than twenty minutes from the cozy suburb where I spent my childhood exists the line—now a series of walls—indicating rich from poor; abundance from desperation; as well as the end of the temperate world and the beginning of the austere, subtropical environs of the Baja Peninsula. It was these two images, or ideas, I believe, that sparked my initial curiosities: the vastness and seemingly wild nature of the landscape, as well as the knowledge that something totally foreign, culturally, and somewhat forbidden, existed so close to my home. And from an early age I knew that I would one day explore both of these realms.

Aside from weekend trips to northern Baja California during the latter part of my teens, my first extended travels to Latin America took place in Costa Rica and consisted primarily of surfing and hiking. By and large these early explorations offered a limited perspective to the life and landscapes of the places I visited. I stayed and worked in American-owned surf camps and spoke limited Spanish, investigated little into local history and culture, and concentrated primarily on finding situations that would allow for a longer stay. Despite my naiveté, however, as well as my ignorance of the complicated situation I was participating in, I do remember feeling that something critical was missing from my consciousness and

N. Einbinder, *Dams, Displacement and Development*, SpringerBriefs in Latin American Studies, DOI 10.1007/978-3-319-51511-3_1

experiences. Eventually my infatuation with natural history overtook my desire to surf, and I began planning trips in order to view birds, rare habitats, and walk through the backcountry. Once I realized the unlimited possibilities for exploring the mountains, forests and countryside, my objectives shifted into an attempt to obtain some form of intimacy within the landscape, as I had in California, and learn all I could about the ecosystems and physical characteristics of the terrain.[1]

My trip to Guatemala, in 2004, fell in the middle of a three-year traveling stint. During this timeframe, I worked and saved (and planned) during part of the year, and spent the little money I earned by exploring the mountains of Latin America or California. Guatemala was part of a four-month loop beginning and ending in Cancún, on Mexico's Yucatán Peninsula. The goal was to experience the region historically and contemporarily inhabited by the Maya people, including southern Mexico, Belize, and Guatemala. For years I had dreamed of visiting this part of Middle America and Mexico, not only for its well-advertised cultural aspects; the colorful Maya villages and ruins of Tikal and Palenque; but also to walk and bus through the volcanic cordilleras and high plateaus of the Sierra, and take inventory of its increasingly rare temperate forests.

Although my awareness regarding the effects of colonialism and the unequal structures of power and 'development' (see Power 2003) expanded significantly while traveling through the Andes the previous year, my experiences in Guatemala brought forth yet another level of intimacy into histories marked by waves of violence and oppression. Particularly apparent were the lasting effects from the civil armed conflict of 1960–1996. Lacking sufficient knowledge of the conflict and the tension that remained, a friend and I planned a three-day walk, by coincidence, through one of the most devastated regions in the country, known as the Ixil Triangle (see Perera 1993; Stoll 1993). Once a stronghold for the Guerrilla Army of the Poor (EGP), an opposition group with wide support from local Ixil, Mam, and Quiche-speaking Mayas, the Guatemalan army targeted the entire region by burning villages of suspected guerrilla supporters, often resulting in complete annihilation (CEH 1999). As pointed out by Guatemalan writer Perera (1993, 62), the number of primarily Indigenous victims was astonishing in the Ixil region during the late 1970s and early 1980s, with over 25,000 inhabitants, or nearly one-third the population, either murdered or disappeared.

Although our walk through the Cuchumatanes was awesome in regard to the landscape (the highest non-volcanic cordillera in Central America), evidence of a lingering violence permeated even the most peaceful and pastoral spaces of the countryside. From one community to the next, we heard stories of massacres, of exile to the forest and into Mexico, and witnessed disturbing levels of poverty,

[1]The idea of becoming immersed within a landscape, issue, struggle, or culture, is an on- going theme in my studies and objectives, which I attribute, at least in part, to my early interest in the work of naturalist and writer Snyder (1996), as well as other 'nature' writers who share the same ideas of 'meditation into place'—a practice and theory which ideally results in a deep, almost spiritual base of knowledge for the natural and human-influenced patterns and processes that make up a landscape.

alcoholism, and environmental degradation. In the highest of villages, over 3000 meters on the stark, moonscape of the altiplano, food appeared scarce, except for cabbage and tortillas, and the men, intoxicated and staggering about in the late day mists, wanted to share with us stories from the war.

"We weren't communists," one man told us, from inside his smoke-filled home. "We farmed potatoes and had some sheep. The army came to kill us all; we left for weeks at a time. They burnt our homes and slaughtered all the animals." In his notebook were written histories—names of the murdered—and drawings of airplanes dropping bombs, and uniformed soldiers with machine guns.

At the time it puzzled me why the army would target such remote and impoverished communities, but equally confusing were the posters of former President General Efraín Ríos Montt—the notorious general dictating the massacres of the early 1980s—tacked up to some of their doors. One man, in the Quiche-speaking village of Xexocom, said that members of Montt's political party, the Guatemalan Republican Front (FRG), showed up before the 2003 election in which Montt was a candidate, and promised money if the people showed support. "They promised all kinds of things," he told me, "and some people believed, even after what happened."

Luckily, Montt did not win the presidential election, but to this day remains a key figure in Guatemalan politics, despite charges of genocide by the United Nations (CEH 1999), and by the Guatemala-based Association for Justice and Reconciliation (AJR) (FHRG 2007).

As written by historical geographer Lovell (1988), the complex patterns embedded within the landscape, culture, and political activities of the Maya peoples can only be interpreted through four and a half centuries of periodic oppression, displacement, and "cycles of conquest"—often carried out as a means for the governing elite to upkeep unequal structures of power and land ownership. Given the rich and varied geography of Guatemala, history has played itself out differently from region to region, often reflecting capitalistic endeavors carried out by the oligarchy and its partners in the United States, and the levels of resistance as employed by the marginalized, Indigenous class. In the Ixil, like so many other highland locations, the production of coffee and the need for cheap, seasonal labor, as well as large tracks of fertile land for whatever crop might be profitable at the time, shaped the resistance movement and impoverishment afflicting the region both historically and into the present day (Perera 1993; Lovell 1995). Although I was still quite unaware of these realities at the time of my initial visit, my somewhat disturbing, yet eye opening experiences in the mountains, as well as my observations from the cities and tourist centers, confirmed the conclusions made by Lovell, and my desire to understand the complexities of such deeply entrenched inequality and dysfunction was well underway.

A fateful meeting with Dr. Catherine Nolin of the University of Northern British Columbia's Geography Department in the Fall of 2007 led to an opportunity that entailed something other than simply 'checking things out' in Latin America, by introducing a plan of formalized study (Master's of Arts—Geography) and investigation into the disturbing and often confusing realities I witnessed on my travels.

During the course of that year my studies focused on Guatemalan politics and history, as well as a critical introduction to development and its role in the making of the 'third world'. A field school to Guatemala that May brought these two elements together, where a series of encounters, not unlike my trip to the Ixil countryside, went deep into the entanglements of global capitalistic development and the enduring conquest as outlined by Lovell (1988) and other scholars (see Perera 1993; Handy 1984; Carmack 1988; Nolin 2006). Through community visits, as well as discussions with key actors in the struggle for environmental and social justice, our delegation witnessed the 'lived experiences' of those affected by foreign-operated mining projects, hydroelectric dams, and severe economic and social repression.

Among these visits included the village of Río Negro, whose historical and contemporary struggles would eventually become the topic for my Master's research, and this book. Aside from pure circumstance, as well as the rare opportunity to have access to such a remote, traditional, and particularly sensitive community,[2] I chose Río Negro and its connection to the Chixoy Dam as an investigative study because of its clear depiction of the relationship between so-called development and violence, as well as the deep historical and ideological divides between the Mayas and more 'westernized' Ladinos.

Not dissimilar to the Indigenous displacements that occurred during the Spanish conquest (Lovell 1988), and continuing into the twentieth century with the advent of export-agriculture such as coffee and bananas (Handy 1984; Schlesinger and Kinzer 1983), the World Bank/Inter-American Development Bank-funded Chixoy Hydroelectric Dam, built between 1976 and 1983, was slated to remove over 3400 Maya-Achí inhabitants from their Chixoy River Valley home, in order to provide 'cheap and reliable' electricity for the minority of citizens with access to services (Witness For Peace 1996). With approximately 800 residents, the village of Río Negro was one of the largest in the basin, occupying a fertile track of land at the rivers' edge, where they grew traditional crops—corn, chiles, beans, and squash—as well as fruits and vegetables for local markets in the nearby centers of Rabinal and San Cristóbal Verapaz. As noted in the limited reports compiled before the dam's construction (Ichon 1978), as well as by previous residents, Río Negro had access to good water, fishing, land, and maintained traditional Achí customs and agriculture not too distant from their pre-conquest ancestors.

Despite World Bank protocol on relocation for development projects (see World Bank 1978; Lynch 2007), Río Negro inhabitants claim that the government organization in charge of the plan, the National Institute of Electrification (INDE), never consulted them, and simply appeared one day by helicopter to inform residents that

[2]Access to Río Negro for our delegation and my future ethnographic research came through Grahame Russell, co-director of Rights Action, a Canadian and U.S. based non-governmental organization (NGO). Since 1994, Grahame has been involved in the struggle for justice in Río Negro, and nearby Rabinal, where many of the survivors of the 1980–83 massacres presently live. Along with Dr. Nolin, Grahame co-led our delegation to various locations in which his organization has affiliation.

a dam would be built several kilometers downstream, and that their village would be flooded. For the first several years of construction, INDE agents made promises regarding compensation—sufficient land, proper homes, and services—but as time passed, it became apparent to residents that no such agreements would be kept (Osorio 2003; Colajacomo 1999). As will be discussed in detail in later chapters, Río Negro's attempt to receive proper compensation for their land, homes, and economic livelihoods, and their eventual decision to collectively resist their pending displacement, was met by targeted assassinations of community leaders, kidnappings, and massacres now confirmed as acts of genocide by the United Nations Truth Commission (CEH 1999, VI, Illustrative Case Annex no. 1). In a series of planned attacks carried out under army orders, over half the residents—men, women, children, and the elderly—were murdered shortly before the dam's final construction. Those who survived the violence fled into nearby mountains, or were taken in as slaves by paramilitaries (Johnston 2005; Witness For Peace 1996; Colajacomo 1999; Osorio 2003).

Not unlike the stories I heard in the Ixil, and confirmed through extensive investigations undertaken by the United Nations (CEH 1999) the Catholic Church (REMHI 1998), and numerous anthropologists (see Carmack 1988), murder and intimidation were tactics frequently employed by President General Ríos Montt, and his predecessor General Roméo Lucas Garcia, in their attempt to eliminate opposition movements throughout the country. Resistance at this time in any shape or form—from unions, churches, human rights, and development groups—was interpreted as guerrilla or communist activity (Manz 1988). And since Río Negro fit this subversive profile, it became one of the 600+ Indigenous villages to be wiped off the map as part of the government's, U.S.-sanctioned, 'scorched earth policy' (CEH 1999).

Most individuals who survived the massacres at Río Negro, and the years of hiding that followed immediately after, presently live with their families in the relocation village of Pacux, the government-built, World Bank-funded relocation 'village' that sparked the initial resistance movement among Río Negro's elders and community leaders. As witnessed during the ethnographic component of my research, life in this cramped, severely impoverished community has all but shattered the more traditional, self-sufficient Achí way of life, as residents must cope daily with the memories of lost mothers and fathers, and attempt to survive in a 'modern' capitalist economy with extremely limited opportunities and resources.

Those few individuals who reached a breaking point in Pacux have since risked their lives in order to move back to the Chixoy River basin and return to a more traditional lifestyle. Located directly above the original Río Negro village site, now submerged by the reservoir, approximately 15 families work the steep, arid slopes, having miraculously succeeded in re-colonizing the land and offering a more self-sufficient existence for their children.

As mentioned, the events from Río Negro and the Chixoy Dam are not uncommon given the historical context of development and modernization as carried by the oligarchy/state, and their foreign, primarily U.S. partners (Jonas 1991). Resultant violence, displacements, and exclusion from these processes have effectively shaped the geography of the country (Lovell 1988), initiated and sustained conflicts (Benson et al. 2008), and shattered Indigenous structures of community and identity (Nolin 2006). Aside from this specifically Guatemalan context,[3] however, lay the powerful role of international actors and the dominant development paradigm existing since the post-WWII era and into the present (Power 2003). Just as the incidents from Chixoy may be viewed as highly symbolic in terms of Guatemalan development and its connections to the state-sponsored violence and conflicts over land, its construction represents the neoliberal development ideology that resulted in thousands of dams and other mega-development projects of that period (and the present), funded and planned by U.S. and European-based financial institutions such as the World Bank, International Monetary Fund (IMF), and the Inter-American Development Bank (IDB) (Witness For Peace 1996; Lynch 2007; McCully 1996). While these projects have done their share of environmental damage, as well as perpetuating dependent relationships between formally colonized nations and institutions of the Global North, they are also responsible for massive worldwide displacements, political unrest, and cultural annihilation (McCully 1996; Escobar 2003; De Wet 2006).

By converging these interrelated topics and interests, my work seeks to illuminate the on-the-ground realities of those affected by this form of neoliberal-inspired development, as it melds into the violent political and historical Guatemalan context. Admitting that the case from Río Negro is unique to the complex nature of Guatemalan politics and the civilian armed conflict, my intention is to utilize the personal histories from this event to bring to light the highly destructive characteristics inherent to this form of development, both in Guatemala and elsewhere, in the past and the present. In addition, I also aim to describe the deep ideological differences that remain in Guatemala between the Maya and Ladino, by illuminating the development perceptions and activities of the Achí survivors of Río Negro.

1.1 Approach

While enjoying the freedom of interdisciplinary work; that is, utilizing key aspects from a range of social science disciplines and theories; the overarching approach for this research is that of a critical, or radical geographic inquiry (see Peet 1977; Aitken and Valentine 2006). As geographer Harvey (2001, 208) states: "A critical geography might go so far as to challenge contemporary forms of political-economic power, marked by hyper-development, spiraling inequalities, and multiple signs of serious environmental degradation." By adopting this critical

[3]Although the aforementioned development activities and violence occur(ed) on Guatemalan soil, most events have international relevance, as investments, ideologies, and power—often in the hands United States citizens and government—have initiated and carried out activities in partnership with dictatorships, and through the use of military enforcements (see LaFeber 1993).

stance, rather than the more traditional position of 'objectivity' and observation, my research seeks to address the urgent social and environmental crises I encountered, by remaining active and questioning of the systems that create and perpetuate injustice and inequality.

Key approaches I use to decipher the events at Chixoy and elsewhere in Guatemala include the spatial unevenness of development (Power 2003), critical North-South connections (Frank 1972), and the consequences of globalization at micro and macro scales (Harvey 2001). Because the Chixoy Dam, as well as other human-displacing projects, are often perceived in the western economy as inevitable in the pursuit of 'progress' (Escobar 2003), I found it necessary to make explicit the inherent flaws of capitalism—what might be interpreted as a Marxist-inspired approach—using the example of Río Negro as a consequence in the most severe case. As stated by geographer Peet (1977, 116):

> The essence of the Marxist argument, therefore, is that inequality is not a "temporary aberration" nor poverty a "surprising paradox" in advanced capitalist societies; instead inequality and poverty are vital to the normal operation of capitalist economies... Unemployment, underemployment, and poverty are inevitably produced by mechanization, automation, and the uneven course of economic development. Inequality underlies our whole economic way of life.

Along with this grounding in critical social theory, my work in Río Negro is aligned with a growing camp of activist anthropologists and geographers dedicated to uncovering and exposing the violence occurring historically and presently in Guatemala (see Nolin 2006; Manz 1988; Carmack 1988; Sanford and Angel-Ajani 2006; Falla 1994). The work of these scholars, including that of Guatemalan anthropologist Myrna Mack (see Lynn 1998), whose documentation of the destruction of rural villages ended with her own assassination by the military in 1990, has been crucial in the dissemination of information often hidden from the public and academic sphere. As repeatedly stressed by anthropologist Manz (2008), there is no choice for the social scientist or ethnographer, in a place such as Guatemala, to remain neutral and uncommitted to work in the promotion of justice and advocacy. Although I was not present in Guatemala at the height of the genocide as Manz and other scholars were, I understand and adhere to this belief after witnessing the blatant injustice and inequalities in Guatemala, which continue unabated.

1.2 Overview

This book is organized into seven chapters, the first of which is this introduction. In chapter two, I describe the research design and methodology. Following this, in chapter three, I offer a critical overview on the topic of development, bringing to light the theoretical lens guiding this research and its analysis. In chapter four, I give a brief historical sketch of Guatemala, with a focus on Indigenous repression and development activities as carried out by the governing Ladino class as well as their international partners. From this broad historical perspective I move into the Río Negro story, utilizing my own experiences, but also relying heavily on previously written documents and reports. Chapter six contains the heart of my thesis: the analysis from my ethnographic research. Finally, in chapter seven, I summarize my findings and conclude this study.

References

Aitken, S. C., & Valentine, G. (2006). *Approaches to human geography*. London, UK: Sage Publications.

Benson, P., Fischer, E. F., & Thomas, K. (2008). Resocializing suffering: Neoliberalism, accusation, and the sociopolitical context of Guatemala's new violence. *Latin American Perspectives, 35*(5), 35–58.

Carmack, R. M. (Ed.). (1988). *Harvest of violence: The Maya Indians and the Guatemala crisis*. Norman, Oklahoma: University of Oklahoma Press.

Colajacomo, J. (1999). The Chixoy Dam: The Maya Achi' Genocide. The story of forced resettlement. Contributing Paper, World Commission on Dams. Cape Town, South Africa. Accessed 10 Mar 2008. http://www.dams.org/kbase/thematic/tr12.htm

Commission For Historical Clarification (CEH). (1999). *Guatemala: memory of silence*. Report of the Commission for Historical Clarification Conclusions and Recommendations Guatemala, http://hrdata.aaas.org/ceh/report/english/

De Wet, C. J. (2006). *Development-induced displacement: Problems, policies, and people*. New York, NY: Berghahn Books.

Escobar, A. (2003). Displacement, development, and modernity in the Colombian Pacific. *International Social Science Journal, 55*(175), 157–167.

Falla, R. (1994). *Massacres in the Jungle: Ixcán, Guatemala 1975-1982*. New York, NY: Westview Press.

FHRG (Foundation for Human Rights in Guatemala. (2007). Genocide charges against Ríos Montt in Guatemala. Accessed 10 Feb 10 2008. http://www.fhrg.org/mambo/index.php?option=content&task=view&id=679&Itemid=2

Frank, A. G. (1972). The development of underdevelopment. In J. Cockcrost, D. James, A. G. Frank, & D. Johnson (Eds.), *Dependence and underdevelopment*. New York, NY: Anchor Books.

Handy, J. (1984). *Gift of the devil*. Toronto, ON: Between The Lines.

Harvey, D. (2001). *Spaces of capital: Towards a critical geography*. London, UK: Routledge.

Ichon, A. (1978). *La Cuenca Media Del Río Chixoy* (Guatemala). Cuadernos de estudio Guatemaltecos 3. Guatemala City, Guatemala.

Johnston, B. R. (2005). Volume one: Chixoy Dam legacy issues study, executive summary: Consequential damages and reparation: Recommendations for remedy. Center for Political Ecology, Santa Cruz, California. http://www.centerforpoliticalecology.org/chixoy.html

Jonas, S. (1991). *The battle for Guatemala: Rebels, death squads and U.S. power*. New York, NY: Westview Press.

LaFeber, W. (1993). *Inevitable revolutions: The United States in Central America*. New York, NY: W.W. Norton & Company.

Lovell, W. G. (1988). Surviving conquest: The Maya of Guatemala in historical perspective. *Latin American Research Review, 23*(2), 25–57.

Lovell, W. G. (1995). *A beauty that hurts: Life and death in Guatemala*. Austin TX: University of Texas Press.

Lynch, B. D. (2007). The Chixoy Dam and the Achi Maya. Violence, ignorance, and the politics of blame. Einaudi Center Working Paper. Cornell University. http://www.economyandsociety.org/events/conf_ethicsglobalization.shtml

Lynn, S. (1998). *The Myrna Mack case: An update*. Washington, DC: National Academic Press.

Manz, B. (1988). *Refugees of a hidden war: The aftermath of counterinsurgency in Guatemala*. New York, NY: SUNY Press.

Manz, B. (2008). The continuum of violence in post-war Guatemala. In Alisse Waterston (Ed.), *The anthropology of war: Views from the frontline*. Oxford UK: Berghahn Books.

McCully, P. (1996). *Silenced rivers: The ecology and politics of large dams*. London, UK: Zed Books.

Nolin, Catherine. (2006). *Transnational ruptures: Gender and forced migration*. Surrey, UK: Ashgate Publishing Ltd.

Osorio, J. T. (2003). *The Río Negro massacres*. Toronto, ON: Jesús Tecú Osorio & Rights Action.

Peet, R. (1977). *Radical geography: Alternative viewpoints on alternative social issues*. London, UK: Taylor & Francis.

Perera, V. (1993). *Unfinished conquest*. Los Angeles, CA: University of California Press.

Power, M. (2003). *Rethinking development geographies*. London, UK: Routledge.

REMHI, Recuperación de la Memoria Histórica - Informe Proyecto Interdiocesano. (1998). *Guatemala: Nunca Más*. ODHAG (Oficina de Derechos Humanos del Arzobispado de Guatemala): Guatemala.

Sanford, V., & Angel-Ajani, A. (Eds.). (2006). *Engaged observer: Anthropology, advocacy, and activism*. New Brunswick, NJ: Rutgers University Press.

Schlesinger, S., & Kinzer, S. (1983). *Bitter fruit: The untold story of the American coup in Guatemala*. Garden City, NY: Anchor Books.

Snyder, G. (1996). *A place in space: Ethics, aesthetics, and watersheds*. Los Angeles, CA: Counterpoint Press.

Stoll, D. (1993). *Between two armies in the Ixil towns of Guatemala*. New York, NY: Columbia University Press.

Witness For Peace. (1996). *A people dammed: The impact of the World Bank Chixoy Hydroelectric Project in Guatemala*. Washington D.C., USA: A Witness for Peace Publication.

World Bank (1978). Staff appraisal report, Guatemala, Chixoy Power Project. Regional Project Department, Latin America and the Caribbean Regional Office. http://web.worldbank.org/external/projects/main?pagePK=64283627&piPK=73230&theSitePK=40941&menuPK=228424&Projectid=P007189

Chapter 2
Research Design

Abstract As stated in the previous chapter, this research attempts to illustrate the 'lived experiences' of those affected by the Chixoy Hydroelectric Dam, and the violence that occurred during and after its construction. In order to achieve this objective I used a range of qualitative research methods, including semi-structured interviews, or *testimonios*, observation, and participation in the 2008 UNBC Geography field school. In addition to the ethnographic component my research utilized secondary documents, which led to a wide-ranging literature review on the topic.

Keywords Human geography · Ethnography · Research design · *Testimonio*

2.1 'The Field'

The bulk of my research was carried out in the field, employing ethnographic methods (Hay 2005; Patton 2002). According to Johnston et al. (2000, 238), ethnography is used "to convey the inner life and texture of a particular social group or locality." Through interviews, observation, and participation in meetings and daily activities, I supplemented what had already been written about the topic, while bringing to light the on-the-ground realities so often left out of development studies.

Ideally, ethnographic research is carried out over an extended period of time, where one can fully engage with a community or particular location. As stated by geographer Taylor (2007, 182), "It is only through time in a community, however, that a story of past and present life emerges." Aside from my preparation in Canada I spent a total of four months in Guatemala, with roughly half of this time in the communities of Rabinal and Río Negro. Although this provided enough time, I believe, to interview a sufficient number of participants to reach saturation (Kirby and McKenna 1989, 123) on certain key themes in my research, as well as achieve a general understanding of what life is like in the communities, I cannot claim in this

N. Einbinder, *Dams, Displacement and Development*, SpringerBriefs in Latin American Studies, DOI 10.1007/978-3-319-51511-3_2

limited time to have a completely in-depth, first-hand knowledge of the everyday realities of living in resettlement and extreme poverty, and believe it would take a much longer commitment to achieve that purpose.

As a result of my commitment to solidarity with those individuals affected by the Chixoy dam and resultant massacres, bias and subjectivity inevitably shaped the research design and approach. My goal, as explicitly stated at various points throughout this study, as well as to all those involved in my research, is to document the effects of the Chixoy Dam on the original residents of Río Negro and their families. As an open critic of mega-projects—and genocide—it was my aim to generate evidence, in the form of oral histories and my own personal observation, that would denounce and expose the institutions and ideologies responsible for the displacement, cultural annihilation, and murder that took place against the people of Río Negro (Witness For Peace 1996; Johnston 2005). Purposefully, I did not interview, or even try to contact, people in support of the dam. If desired, I potentially could have located individuals involved in the paramilitary at the time of the counterinsurgency, and heard, first-hand, their side of the story. Although this may have provided a more balanced study, I believe that by operating in this manner I would have jeopardized my relationship and trust with survivors, as well as broken my commitment, and solidarity, with respondents (see Nolin 2006, 10–18). And therefore I chose to omit this possibility entirely.

As mentioned in the introduction, my fieldwork for this research began while participating in the UNBC field school during the month of May 2008. Co-taught by Dr. Catherine Nolin and Grahame Russell of Rights Action, this course exposed a handful of students, and myself, to the realities of Indigenous displacement, the legacies of the 36-year internal armed conflict, and the current struggles for rights-based development. Among so many localities, our group arrived to Rabinal, as well as the remote resettlement community of Río Negro. Despite my previous knowledge of Chixoy, it was at this point that I made initial connections with local leaders and heard, first-hand, the stories that I had read so much about. Sebastian Iboy Osorio, who was president at the time of the local COCODE (community development council) in Río Negro, was our guide for the trip to his community, which included a boat ride across the reservoir, and a night at the recently built Centro Histórico (Historical Center). At various points, Sebastian shared his personal story, or *testimonio*, of the violence that occurred in Río Negro, as well as stories of his life, and that of the entire community. He also spoke of development projects taking place in Río Negro, and guided us up the mountain to the site where he lost most of his family during the March 1982 massacre at Pocoxom.

After seven months in Canada, writing my research proposal and gaining approval from the Research Ethics Board (REB) at UNBC, I returned to Guatemala. Almost immediately, I joined a human rights delegation in Rabinal, with Grahame Russell as well as two professors from the University of Guelph. The purpose of this visit was for Grahame to introduce me to Carlos Chen, who was co-director of COCAHICH (Coordinator of the Communities Affected by the Construction of the Chixoy Dam), as well as a Río Negro massacre survivor. Due to Grahame's tight affiliation and support for the various organizations working towards justice and reparations in Rabinal and Pacux,[1] Carlos, as well as other members, warmly welcomed me and encouraged further contact upon my return. Word was also sent, through these men, to Sebastian in Río Negro, informing him of my objectives to carry out this study. It goes without saying that my proposed work in Rabinal, and Río Negro, would have been much more difficult if not impossible without these initial introductions, and obvious support from Grahame.

2.2 Data Collection

The bulk of my data was obtained through semi-structured, open-ended interviews, as outlined by Dunn (2005) and Leech (2002). According to geographer Dunn (2005): "Semi-structured interviews employ an interview guide. The questions asked in the interview are focused and deal with the issues or areas judged by the researcher to be relevant to the research question" (88). These "interview/conversations" (Nolin 2006, 16) were typically long (one to three hours), exhaustive on the particular issues, and highly emotional. Although interviews were highly variable, based on the age of the respondent, or whether they were a current resident of Pacux or Río Negro, I followed a standard set of questions.

This form of in-depth interviewing, also known in geographic methodologies as *testimonio*, has demonstrated, according to ethnographer Haig-Brown (2003), "an apt approach to respectful, useful, and overtly political research" (416). Haig-Brown (2003) also states that the *testimonio* "has the potential to create space for other impossible knowledges that are underrepresented or invisible within conventional academic discourses" (416). The *testimonio* as a research method is also a useful way to obtain the histories of a community without interviewing every single person, a concept Hanlon and Finola (2000, 267) refer to as "collective remembrance." As stated by Haig-Brown (2003, 420): "Central to the testimonial is the fact that the life story presented is not simply a personal matter; rather, it is the story of an individual who is also a part of a community. A *testimonio* presents the life of a person whose experiences, while unique, extend beyond her/him to represent the group of which she/he is a member."

While hoping for the most accurate depiction of what took place in Río Negro at the time of the construction of the Chixoy Dam, I do realize the concern over accuracy surrounding the use of in-depth interviews, or *testimonio*, and concur with Clifford and Marcus (1986, 7) in that all, "ethnographic truths are thus inherently

---

[1]A handful of local Maya-Achí groups were founded in the early 1990s to address the problems associated with the Chixoy dam. COCAHICH, and ADIVIMA (Association For the Integral Development of the Victims of Violence in the Verapaces, Maya-Achí) are a couple of the most well known.

partial—committed and incomplete." As stated by Nolin (2006, 18), "in-depth interviewing can be an excellent tool if the researcher is flexible enough to allow for some inconsistency, repetition, exaggeration, or omission in order to understand the subject of their research project in more depth." She also goes on to mention that "when the limitations of *testimonio* and in-depth interviewing are understood and acknowledged, then the identified shortcomings can be dealt with through supplementation with secondary material such as newspapers, journalistic accounts and other written material" (Nolin 2006, 18).

Following the interviews of several key informants (see Patton 2002, 321), such as Sebastian in Río Negro, and Carlos in Rabinal, a snowball sampling process (Monk and Bedford 2005, 62) led me to other interviewees. My initial goal was to collect at least ten testimonios, from a diverse set of individuals—male, female, older, younger, from Pacux and Río Negro—as well as community leaders who were involved in the social justice movement, and could speak of issues that affected the entire community. In the end, I conducted 15 interviews. This number, I believe, provided a sufficient amount of data necessary to supplement previously written accounts, while also allowing for the development of new themes. Although I do not claim to have the 'whole' story collected through my limited amount of interviews and time spent among residents, the 15 interviews illuminated several themes to be discussed in chapters 5 and 6.

Depending on whether I was in Pacux or Río Negro, or if other outsiders were visiting, I had variable success in locating respondents willing to speak with me. The collection of some of the richest data occurred during discussions and lectures given to visiting groups while I was in Río Negro. Following these events I was usually able to ask questions that pertained to my research, as well as get contact information for future interviews. In Pacux, Heidi McKinnon, a volunteer with The Advocacy Project, was immensely helpful in leading me to respondents, particularly women, which in turn resulted in other interviews. Although I am content with the number of interviews I conducted in Pacux, I could have conducted many more if needed. Residents were particularly fond of Heidi and her work in the community, and took me in with an incredible amount of confidence and trust.

Interviews generally took place in the homes of individuals, on the deck of the Centro Histórico in Río Negro, or at the ADIVIMA office in Rabinal. I used a digital tape recorder and asked everyone's permission before turning it on. No one requested to be anonymous nor had a problem with the recording of our discussions. Due to the highly publicized nature of the Río Negro massacres, and the success they have had with international solidarity groups, survivor's expressed no fear in giving *testimonios* or participating in studies and media events. Many respondents asked to have their pictures taken and included in the report with their full names used.

As a strategy to obtain the most accurate information possible, as well as ask newly emerging questions, I interviewed several key informants, such as

SEBASTIAN[2] from Río Negro, as well as MARIO from Pacux, on multiple occasions (Baxter and Eyles 1997, 5). Because of the informal nature of my visit, especially in Río Negro, many discussions also took place while walking, eating, or in the boat, which I wrote about in my notebook.

Participant observation was another aspect of my fieldwork, which generated a sizeable amount of data both in the communities, as well as throughout my time elsewhere in Guatemala. As stated by Kearns (2005, 196), "participant observation of a geographer involves strategically placing oneself in situations in which systematic understandings of place are most likely to arise." Although interviewing key individuals was essential in gathering data about the past and present conditions for the lives of the survivors, my observations and conversations in an everyday, unstructured setting offered new perspectives crucial for this study. While in communities, I always had a small notebook with me, and would wait for appropriate times to jot down notes. I then transferred all the notes into a larger notebook, which I used in my analysis.

2.3 Analysis

Data analysis followed a systematic protocol under the guidance of Kirby and McKenna (1989), Patton (2002), Hay (2005) as well as my supervisor, Dr. Nolin. This entailed transcription of interviews from digital to word documents, followed by an identification of recurring themes through a coding process. The most pronounced themes I found within the interviews were: (1) Effects of Violence; (2) Economic Impacts; and (3) Contemporary Development. After identifying these initial over-arching themes, I identified subthemes that corresponded, and supported, the major topics. For the analysis chapter of this book, I synthesized the themes and subthemes into a narrative. In this narrative I also included direct quotes of selected respondents, in order to strengthen the weight of my argument (Baxter and Eyles 1997).

2.4 Ethical Considerations

At the forefront of my ethical concerns was that of representation. As stated by anthropologists Marcus and Fischer (1999, 83), "the crisis [of representation] arises from uncertainty about adequate means of describing social reality." At various points throughout the research I caught myself thinking: *Who am I, a North*

[2]All interviewees involved in this study are mentioned in CAPITAL letters, in order for the reader to make a distinction between them and other individuals. It should also be noted that, due to the sensitive material some interviewees shared with me, I decided to change several names into pseudonyms—despite the fact that no one asked to remain anonymous.

American white male, to interpret the stories from Río Negro, to express their dreams of development, while molding it into complex theories and explanations? In grappling with this dilemma, I attempted to remain conscious of my limitations as an interpreter of knowledge and experiences, and kept in mind that all writings of this nature (whether the author believes so or not) are highly dependent on the individual perceptions and background of the researcher. I also attempted to remain explicit and self-reflective (Dowling 2005, 28) throughout the process, from research to writing to presentation, with the full acknowledgement that what I saw, wrote about, and questioned, was entirely based on my emotions, as well as my academic background and interests.

Alongside the issue of representation and interpretation is that of 'positionality' (see Aitken and Valentine 2006; Sundberg 2003). As explained by England (2006, 289):

> Positionality is about how people view the world from different embodied locations. The situatedness of knowledge means whether we are researchers or participants, we are differently situated by our social, intellectual and spatial locations, by our intellectual history and our lived experiences, all of which shape our understandings of the world and the knowledge we produce…Positionality also refers to how we are positioned (by ourselves, by others, by particular discourses) in relation to multiple, relational social processes of difference (gender, class, 'race/ethnicity', age, sexuality, and so on), which also means we are differently positioned in hierarchies of power and privilege.

As a North American (with all the privileges and power that go along with it) I often wondered what interviewees thought about me, and whether there was any resentment—although none was visibly expressed. I also wondered, at times, whether my respondents—especially in Pacux—hoped that by speaking with me, they might be able to get some form of financial gain from it. There was one instance where an interviewee talked at length about his financial woes, and how he hoped that Rights Action would help fund some of his project ideas. He then went on to mention, several times, my relationship with Grahame Russell. This was an isolated incident, and for the most part I firmly believe that respondents knew my objectives well, respected me, and appreciated my solidarity to their cause.

As stated earlier, my own background of education, upbringing, and political stance, as well as my position of solidarity and compassion for those I interviewed affected the way I wrote and planned this study. At times, my own attachment to the survivors, as well as the emotional challenges I had in dealing with the stories I heard made it difficult to maintain my 'professionalism' (Hume and Mulcock 2004, 154). I felt compelled to help respondents any way that I could and at times held myself back from giving some of the women in Pacux personal donations after hearing about their struggles. As a means of dealing with these emotions, while also desiring to achieve my research objectives, I attempted to study my own emotions by keeping notes and opening up to survivors as a friend. This, in turn, empowered and propelled me, and made my small and seemingly insignificant objectives feel worthy.

Despite the consideration of all these issues, I believe that I acted in an ethical manner throughout this project. I worked in collaboration with the participants of the study, and tried my best to employ methods that would empower and relate their story, in the most accurate way possible, to an audience that might not be reached were it not for someone with the privilege to listen and observe.

References

Aitken, S. C., & Valentine, G. (2006). *Approaches to human geography*. London, UK: Sage Publications.

Baxter, J., & Eyles, J. (1997). Evaluating qualitative research in social geography: Establishing 'rigour' in interview analysis. *Transactions of the Institute of British Geographers, 22*, 505–525.

Behar, R. (1997). *The vulnerable observer: Anthropology that breaks your heart*. Boston, Ma: Beacon Press.

Bradshaw, M., & Stratford, E. (2005). Qualitative research design and rigour. In I. Hay (Ed.), *Qualitative research methods in human geography*. Oxford, UK: Oxford University Press.

Clifford, J., & Marcus, G. E. (Eds.). (1986). *Writing culture: The poetics and politics of ethnography*. Los Angeles, CA: University of California Press.

Dowling, R. (2005). Power, subjectivity, and ethics in qualitative research. In Iain Hay (Ed.), *Qualitative research methods in human geography*. Oxford UK: Oxford University Press.

Dunn, K. (2005). Interviewing. In I. Hay (Ed.), *Qualitative research methods in human geography*. Oxford, UK: Oxford University Press.

England, K. (2006). Producing feminist geographies: Theories, methodologies, and research strategies. In S. C. Aitken & G. Valentine (Eds.), *Approaches to human geography* (pp. 286–297). Los Angeles, CA: Sage Press.

Haig-Brown, C. (2003). Creating spaces: Testimonio, impossible knowledge and academe. *Qualitative Studies in Education, 16*(3), 415–433.

Hanlon, C. N., & Finola, S. (2000). Gendered spaces of terror and assault: The testimonio of REMHI and the Commission for Historical Clarification in Guatemala. *Gender, Place and Culture, 7*(3), 265–286.

Hay, I. (Ed.). (2005). *Qualitative research methods in human geography*. Oxford UK: Oxford University Press.

Hume, L., & Mulcock, J. (2004). *Anthropologists in the field: Cases in participant observation*. New York, NY: Columbia University Press.

Johnston, R. J., Gregory, D., Pratt, G., & Watts, M. (2000). *The dictionary of human geography*. New York, NY: Blackwell Publishers.

Johnston, B. R. (2005). Volume one: Chixoy Dam legacy issues study, executive summary: Consequential damages and reparation: Recommendations for remedy. Center for Political Ecology, Santa Cruz, California. Accessed 10 Mar 2008. http://www.centerforpoliticalecology.org/chixoy.html

Kearns, R. A. (2005). Knowing seeing? Undertaking observational research. In I. Hay (Ed.), *Qualitative research methods in human geography*. Oxford, UK: Oxford University Press.

Kirby, S., & McKenna, K. (1989). *Experience, research, social change: Methods from the margins*. Toronto, ON: Garamond Press.

Leech, B. L. (2002). Asking questions: Techniques for semistructured interviews. *Political Science and Politics*, *35*(4), 665–886 (American Political Science Association).

Marcus, G. E., & Fischer, M. J. (1999). *Anthropology as cultural critique: An experiment*. Chicago, IL: University of Chicago Press.

Monk, J., & Bedford, R. (2005). Writing a compelling research proposal. In Iain Hay (Ed.), *Qualitative research methods in human geography*. Oxford, UK: Oxford University Press.

Nolin, C. (2006). *Transnational ruptures: Gender and forced migration*. Surrey, UK: Ashgate Publishing Ltd.

Patton, M. Q. (2002). *Qualitative research and evaluation methods*. Los Angeles, CA: SAGE Press.

Sundberg, J. (2003). Masculinist epistemologies and the politics of fieldwork in Latin Americanist geography. *The Professional Geographer, 55*(2), 180–190.

Taylor, M. J. (2007). Militarization and the environment. *GeoJournal, 69,* 181–198.

Witness For Peace. (1996). *A people dammed: The impact of the World Bank Chixoy Hydroelectric Project in Guatemala*. Washington D.C., USA: A Witness for Peace Publication.

Chapter 3
Development: A Critical Backgrounder

Abstract This chapter provides a condensed review on economic development and globalization, as well as a short discussion on dams and displacement on a global scale. The point is to offer the reader a better understanding and foundation of the critical position I assume in regard to these topics, while in turn providing an introduction on the contrasting debates surrounding the very meaning and interpretation of development and/or progress.

Keywords Development · Guatemala history · Hydroelectric dams · Capitalism · Progress · Poverty · Indigenous peoples

> When we project modernity, as we define it, as the inevitable destiny of all human societies, we are being disingenuous in the extreme. Indeed, the Western model of development has failed in so many places in good measure because it has been based on the false promise that people who follow its prescriptive dictates will in time achieve the material prosperity enjoyed by a handful of nations of the West… In reality, development for the vast majority of the peoples of the world has been a process in which the individual is torn from his past, propelled into an uncertain future, only to secure a place on the bottom rung of an economic ladder that goes nowhere.
>
> Davis (2009, 196)

The definition of development varies widely, from its traditional western assumption of the production and accumulation of capital (Escobar 1992), to its meaning among certain Indigenous groups—including the Maya of Guatemala—which often relates to a more spiritual view of land and resources, as well as the preservation of customs and livelihoods (Davis 1993; Ashton and Bryan 2002; Amnesty International 2003; Fischer and Brown 1996). While universal definitions of the term are almost always deemed positive, offering solutions to poverty, inequality, and access to services (Loker 1996, 77), a wide range of development scholars (see Rist 2004), as well as a growing number of empirical studies insist that the pro-growth, positivist model dominating the past century completely degrades the kind of development that may better represent marginalized peoples, while truly only benefiting a select minority.

N. Einbinder, *Dams, Displacement and Development*, SpringerBriefs in Latin American Studies, DOI 10.1007/978-3-319-51511-3_3

Free from the constraints of regulation, the present 'neoliberal' development paradigm has led to a dramatic increase in consolidation and power by multinational corporations, which in turn become 'unaccountable to government' (Hong 2000, 8). Projects such as mining (Holt-Giménez 2007), oil extraction (Watts 2004), export-agriculture on an industrial scale (Bello 2009; Handy 1984), and textile production (Keeling 2004, 10), exemplify this model of privatization, modernization, and exclusion, and often result—at least in countries such as Guatemala—in an increase in poverty for the vast majority, as well as, widespread environmental damage (Escobar 1995). In defense of these activities and programs, pro-neoliberal economists often cite quantitative measures, showing improvements in livelihood to those affected; indices that quote macroeconomic growth and consumption habits; yet as Galeano states (1997, 222), "statistics admit, but do not repent. In the last resort, human dignity depends on cost-benefit analysis, and the sacrifice of the poor is nothing but the 'social cost' of Progress." On similar ground, Davis (2009, 196–197) illustrates the realities of how we define 'progress,' often hidden from economic spreadsheets and databases:

> Consider the key indices of the development paradigm. An increase in life expectancy suggests a drop in infant mortality, but reveals nothing of the quality of the lives led by those who survive childhood. Globalization is celebrated with iconic intensity. But what does it really mean? In Bangladesh, garment workers are paid pennies to sell clothing that retails in the United States and Canada for tens of dollars. Eighty percent of the toys and sporting goods sold in America are produced in sweatshops in China, where millions work for wages as low as 12 cents an hour, 400,000 die prematurely each year due to air pollution, and 400 million people do not have access to potable water, so ruined are the rivers with industrial wealth…

Explanations for the current paradigm are often linked to the post-WWII era, at the beginning of the Cold War (Ziai 2007). The rise of the U.S. power, anti-communist policies and the creation of the World Bank and the IMF were essential in putting modernization and global capitalism to work in the Global South, thus maintaining dependent relationships between former imperial colonies and Western societies (Frank 1972). Through its massive lending programs—resulting in some of the world's largest dams and industrial agricultural systems—these international financing agencies created a system of debt and vulnerability for receiving nations, while effectively keeping markets open for the continuation of exploitation and increasingly consumptive appetites in the Global North[1] (Hong 2000, 7–11; O'Brien 2007). As stated by food and land policy analyst Holt-Giménez (2007, 5), the World Bank—despite its stated objective of poverty reduction and reform—works primarily as an agent to restructure space and politics

[1]As an element of Structural Adjustment Programs (SAP), the World Bank loans out capital based on the agreement that certain reforms, such as trade liberalization and privatization, become instituted by receiving governments. Typically, this makes it easier for First World entrepreneurs to exploit with greater freedom. Critics, such as Bello (2009), argue that these programs increase poverty rather than reduce it, by perpetuating neo-colonial policies of unequal power relations and distribution of land.

within the Global South to allow the entrance of foreign capitalists. For a large percentage of Indigenous and marginalized peoples these policies have perpetuated historic struggles and begun new ones in their wake, as land and tradition must be continually defended (Bodley 1990).

As will be described at length in the following chapter, capitalist development in Guatemala is marked by deep inequalities in political power and access to land and resources (Lovell 1988; CEH 1999). Aside from the momentary progressive gains made between 1944 and 1954 during the democratically elected administrations of Presidents Juan José Arrévalo and Jacobo Arbenz Guzman, which included agrarian reform, access to social services, and increased rights to workers (Chasteen 2006), present-day Guatemala suffers from extreme economic and social inequalities (Kurtenbach 2008), effectively maintaining the highest Human Poverty Index rating in all the Americas, after Haiti (Centre for Economic and Social Rights 2008, 2).

According to Taylor et al. (2006), one may view the Guatemalan development policy in its relation to the land-holding elite, and their need for cheap labor by the peasant majority. Along with this oppressive (post) colonial structure, deep historic divides between ideologies and how development is carried out exist throughout the country, often leading to violent popular uprisings and reprisals by the state, as well as persistent rural and urban poverty, and high rates of migration (May 1999; Manz 2005; Taylor et al. 2006; Nolin 2006). As stated by Maya activist Raxche' (1996, 77):

> In Guatemala development plans are complicated by the existence of two major peoples, markedly different in history, language, and culture and with a history of unequal power relations. In this context, growth, advancement, excellence, improvement, and change are relative; what direction does 'development' have to take in order to qualify as advancement, positive change, or improvement if what the Ladino cultural community considers advancement is the continuation of colonial subjugation and domination of the Maya cultural community? Development philosophies are not formulated by the Maya, who, because of the cultural oppression they have suffered for five hundred years, have their own concept of the type of development that the country needs.

Along with much of the Indigenous populations around the world, the Maya campesinos, or traditional agriculturalists of Guatemala, have suffered disproportionately from displacement and exclusion (Munck 2005, 112; Ashton and Bryan 2002), as modern capitalist development often discredits their 'anti-modernist' ideologies and livelihoods (Holden and Jacobson 2008; Loker 2006). According to cultural anthropologist Davis (1993), a fundamental difference exists in how Indigenous peoples view land and 'development,' which in turn leaves them more vulnerable to the negative effects of globalization. As stated by Davis (1993, 11):

> Indigenous peoples—in contrast to Western economists and development planners—do not view land as a 'commodity' which can be bought and sold in impersonal markets, nor do they view the trees, plants, animals, and fish which cohabit the land as 'natural resources' which produce profits or rents. To the contrary, the indigenous view—which was probably shared by our ancestors prior to the rise of the modern industrial market economy—is that land is a substance endowed with sacred meanings, embedded in social relations, and fundamental to the definition of people's existence and identity.

Generally speaking, the Maya are noted to hold similar values to land (Lovell 1988; Oxfam 2005) and how it is developed, contrasting sharply from development activities pursued, and supported by the state (Jonas 1991, 2000; Kurtenbach 2008; World Bank 2008). In a recent study addressing the issue of mining in the country—one of the most controversial development activities to date—Holden and Jacobson (2008, 135) found that "activities such as open pit mining are utterly alien to the Mayan cosmovision," and that its regional opposition "is another demonstration of resistance to neoliberalism" (145). Holden and Jacobson (2008) also go on to assert that this form of capitalist development threatens the age-old cultural practices endemic to the Maya-Mam of San Marcos (where the study took place), due to involuntary displacement, and the changing of patterns inherent to their traditional livelihood. According to the Minority Rights Group International (2008a), the Guatemalan government admitted that certain linguistic groups such as the Chorti, Mam, Chuj, and Sipakapense "will be faced with ethnocide if the mining projects are not handled appropriately." In another study, geographers Steinberg and Taylor (2008), find similar cultural deterioration in the Cuchumatan Highlands of Guatemala, where they indicate that globalization, as well as population pressures, is resulting in negative pressures on maize diversity and traditional agro-ecological knowledge.

Aside from these contemporary and localized examples, the issue of land, and how it is perceived and utilized, has been an on-going struggle between the Maya majority and Ladino landholders and politicians since the arrival of the Spanish nearly 500 years ago (Grandin 2004; Lovell 1988; Fischer and Brown 1997). Whether it's mining, industrial export-agriculture, or the building of dams, the development model expressed in Guatemala displaces and discredits the livelihood of rural Mayas, and offers little respite to the debilitating poverty that afflicts much of the nation (Kurtenbach 2008). And yet despite these well-documented realities, the World Bank continues to insist on a highly debatable scheme of modernization and free-market reforms (Gauster and Isakson 2007; World Bank 2004; Holt-Giménez 2007), and the government, whether dictatorships or democratically elected, remain predominantly compliant to the Bank's prescriptions.

3.1 'Post' or 'Rights-Based' Development

Challenges to these ever so trenchant systems of economics and government come from both North and South, and a growing resentment is evident among many global citizens towards the dominant model (Loker 1997, 76; Escobar 1992; Klein 2007; Wright 2004; Simon 2003). Postdevelopment is one of the recent theories that critique "western notions and assumptions of superiority and expertise that are seen to very often accompany development interventions and aid" (Sidaway 2007, 347). According to geographers Peet and Hartwick (1999, 11), the framework of postdevelopment suggests that we completely redefine what is meant by 'progress' and 'modernity.'

From a Latin American perspective, Escobar (1992, 22) insists that within growing awareness of the post-WWII "hegemony of development"—which, in its part, has contributed to the creation of the 'Third World' through its poverty, conflict, and debt—movements at the grassroots level, as well as scholarly opposition, will aid the postdevelopment momentum already in place. Guatemala, among other Latin American nations, provides a clear example of local opposition movements towards the dominant development model, as well as communities embracing 'alternative' and 'sustainable' development activities as a means to combat poverty and environmental degradation, and maintain their culture (Holden and Jacobson 2008; International Rivers 2008; COPAE 2006; Warren 1998). Since the signing of the Peace Accords in 1996, hundreds, if not thousands, of local and national development groups have formed, or reorganized across the country. Organizations such as the National Indigenous and Peasant Coalition (CONIC), Campesino Committee of the Highlands (CCDA), and Madre Selva have been particularly active and influential in their work for environmental justice, the enforcement of fair trade practices, agrarian reform (as promised by the Accords), the revitalization of language and culture, and against imposed development projects that perpetuate historical struggles (Viscidi 2004; Montejo 2005).

In Rabinal and Río Negro, the specific locations for this study, local Maya-Achí development and human rights groups are actively pursuing what might be deemed a 'rights-based' development initiative, which according to the Washington Office on Latin America (2003, 2), is described as a system where "people have a right to participate in the political and policy decisions that shape their lives." These initiatives include sustainable agriculture (in contrast to export based), and educational opportunities that focus on Indigenous empowerment and learning skills that are applied directly to their own communities. Specific development projects, such as the Centro Historico y Educativo at Río Negro, which brings income to the entire community while educating international and national students, will be thoroughly discussed in the analysis and discussion chapters of this thesis.

3.2 Dams and Displacement

A critical discussion on development, particularly in a case involving a large hydroelectric dam, would be incomplete without touching on the issue of involuntary human displacement. Despite what we know about displacement, as a result of conflict and war, we have yet to calibrate the number of people who are displaced, often forcibly, due to large-scale infrastructure projects, particularly in 'third world' nations (IDMC 2008). Although involuntary human displacement results from a wide range of development activities—such as mining, industrial agro-business, super-highways, tourism, and conservation—the 45,000+ large dams built worldwide for irrigation and hydroelectric power contribute significantly to that number, with as many as 80 million people displaced in the last century (Stanley 2004, 8–9). As symbols of modernity, national pride, and domination over the landscape, large dams are highly emblematic of the post-WWII development model, and have been historically and presently supported by the World Bank and other international financial institutions (McCully 1996). As McCully states in his landmark book, Silenced Rivers (1996, 237):

> Perhaps more than any other technology, massive dams symbolize the progress of humanity from a life ruled by nature and superstition to one where nature is ruled by science, and superstition vanquished by rule of rationality. They also symbolize the might of the state that built them, making huge dams a favorite of nation-builders and autocrats.

Although dam-generated energy is often sold to the public as 'environmentally friendly,' the resulting damage to ecosystems and displaced communities is staggering (Witness For Peace 1996, 29), and well hidden by repressive governments, as was the case with Chixoy, among many others. As commented by anthropologist Thayer Scudder, in his discussion on the myths of clean, green hydroelectric energy (2005):

> Large dams are flawed for many reasons. Benefits are overstated and costs understated. Especially serious are the adverse environmental impacts on the world's river basins, impacts that tend to be irreversible where dams are built on mainstreams and large tributaries. Implementation continues to impoverish the majority of those who must be resettled from reservoir basins and project works, and to adversely affect millions of people who live below dams whose living standards are dependent on natural food regimes.

Relevant to the Chixoy case is the fact that a disproportionately high percentage of those affected by dam development worldwide are of ethnic minorities and Indigenous descent, thus adding to the argument that, "the creation of a global economy and its accompanying industrial infrastructure has only been possible through dismantling local cultures and reassembling them in more modern forms" (Colchester 2000, 43). As stated by geographers Ashton and Bryan (2002, 4) in their report on Indigenous peoples and the environment: "Regardless of whether state's formally recognize indigenous peoples' use and occupancy of land and resources, development projects, and privatization of land ownership introduce dramatic changes in by indigenous peoples' societies." These 'dramatic changes,' as outlined through countless books and case studies (see McDowell 1996; Stanley 2004; De Wet 2006; Witness For Peace 1996; Windsor and McVey 2005) as well as what I learned first-hand regarding the Chixoy Dam, prove to be highly disruptive to Indigenous and non-Indigenous peoples alike, and too commonly result in moderate to severe socioeconomic problems, such as increased drug and alcohol abuse, extreme poverty, cultural annihilation, and disease. Despite this growing body of knowledge, however, countries such as Guatemala still pursue development projects that displace and dismantle communities (McGahan and Hilly 2008; Rights Action 2006). One such government-initiated project, the proposed Xalala Dam in Northern Guatemala, would flood 36 villages if completed (McGahan and Hilly 2008). Widespread local opposition, however, supported by international solidarity groups and national non-governmental organizations (NGOs) have effectively blocked the project for the time being, and with the prospects of another Chixoy disaster, even the World Bank has backed away from funding the project.

While brief, the hope of this summary is to offer the reader a glimpse of my theoretical framework, as well as introduce the present literature and debates. In the following chapter I will delve deeper into the Guatemalan context, offering a historical overview with special attention to the very notion of development, and the complex entanglements between government and economic growth.

References

Amnesty International. (2003). *Guatemala: Fear for safety and death threats*. Urgent Action, Amnesty International.

Ashton, A., & Bryan, J. (2002). *Indigenous peoples and the environment. Institute of international studies*. Berkeley: University of California.

Bello, W. (2009). *The food wars*. London, UK: Verso.

Bodley, J. H. (1990). *Victims of progress* (3rd ed.). Los Angeles, California: Mayfield Publishing Company.

Centre for Economic and Social Rights (CESR) and Instituto Centroamericano de Estudios Fiscales (ICEFI). (2008). Economic, social and cultural rights in Guatemala: A selective briefing. Submitted to the Office of the High Commissioner for Human Rights on the occasion of the second session of the Universal Periodic Review May 2008. www.cesr.org/…/UPR%20Guatemala%20submission%202008.pdf

Chapman, J. (2003). *Rights-based development: The challenge of change and power*. Global Poverty Research Group. www.gprg.org

Chasteen, J. C. (2006). *Born in blood and fire: A concise history of Latin America*. New York, NY: W.W. Norton and Company.

Colchester, M. (2000). Dams, indigenous peoples and ethnic minorities. World Commission on Dams Thematic Reviews. Social issues I.2. Accessed April 10, 2008. www.dams.org/kbase/thematic/tr12.htm

Commission For Historical Clarification (CEH). (1999). *Guatemala: Memory of silence. Report of the commission for historical clarification conclusions and recommendations Guatemala*, http://hrdata.aaas.org/ceh/report/english/

COPAE (Pastoral Commission for Peace and Ecology). (2006). Open doors to resource extraction. In *Guatemala Human Rights Commission-USA (GHRC), DR-CAFTA, effects and alternatives* (pp. 27–30). http://alainet.org/active/29034

Davis, S. H. (Ed.) (1993). *Indigenous views of land and the environment*. World Bank discussion paper. World Bank, Washington DC, USA.

De Wet, C. J. (2006). *Development-induced displacement: Problems, policies, and people*. New York, NY: Berghahn Books.

Escobar, A. (1992). Imagining a post-development ear? Critical thought, development and social movements. *Social Text*, 31–32.

Escobar, A. (1995). *Encountering development: The making and unmaking of the Third World*. Princeton, NJ: Princeton University Press.

Fischer, E. F., & Brown, R. M. (1996). *Maya cultural activism in Guatemala*. Austin, TX: University of Texas Press.

Frank, A. G. (1972). The development of underdevelopment. In J. D. Cockcrost, A. G. Frank, & D. Johnson (Eds.), *Dependence and underdevelopment*. New York, NY: Anchor Books.

Galeano, E. (1997). To be like them. In M. Rahnema & V. Bawtree (Eds.), *The post-development reader* (pp. 214–222). London, UK: Zed Books.

Gauster, S., & Isakson, S. R. (2007). Eliminating market distortions, perpetuating rural inequality: An evaluation of market-assisted land reform in Guatemala. *Third World Quarterly, 28*(8), 1519–1536.

Grandin, G. (2004). *The last colonial massacre: Latin America in the Cold War.*
Handy, J. (1984). *Gift of the devil.* Toronto, ON: Between The Lines.
Holden, W. N., & Jacobson, R. D. (2008). Civil society opposition to nonferrous metals mining in Guatemala. *Voluntas, 19*, 325–350.
Holt-Giménez, E. (2007). Land-gold-reform: The territorial restructuring of Guatemala's highlands. Development report number 16. Food First Institute for food and development policy. Oakland, California.
Hong, E. (2000). *Globalisation and the impact on health: A third world view.* Third World Network. Accessed May 5, 2009 http://www.phmovement.org/pubs/issuepapers/hong.html
IDMC (Internal Displacement Monitoring Centre). (2008). Internal displacement: Global overview of trends and developments in 2008.
International Rivers. (2008). *Financial crisis stops Xalala Dam—For now.* Berkeley California: International Rivers. Accessed May 5, 2009. http://www.internationalrivers.org/node/3538
Jonas, S. (1991). *The battle for Guatemala: Rebels, death squads and U.S. power.* New York, NY: Westview Press.
Keeling, D. J. (2004). Latin American development and the globalization imperative: New directions, familiar crises. *Journal of Latin American Geography, 3*(1), 1–21.
Klein, N. (2007). *The shock doctrine.* New York, NY: Metropolitan Books.
Kurtenbach, S. (2008). *Guatemala's post-war development: The structural failure of low intensity peace.* Project Working Paper No. 3. Institute for Development and Peace (INEF). University of Duisburg-Essen, Germany.
LaFaber, W. (1993). *Inevitable revolutions: The United States in Central America.* New York, NY: W.W. Norton & Company.
Loker, William M. (1996). "Campesinos" and the crisis of modernization in Latin America. *Journal of Political Ecology, 3,* 69–88.
Lovell, W. G. (1988). Surviving conquest: The Maya of Guatemala in historical perspective. *Latin American Research Review, 23*(2), 25–57.
Manz, B. (2005). *Paradise in ashes: A Guatemalan journey of courage, terror, and hope.* Los Angeles, CA: University of California Press.
May, R. A. (1999). "Surviving all the changes is your destiny": Violence and popular movements in Guatemala. *Latin American Perspectives, 26,* 68–91.
McCully, P. (1996). *Silenced rivers: The ecology and politics of large dams.* London, UK: Zed Books.
McDowell, C. (Ed.). (1996). *Understanding impoverishment: The consequences of development-induced displacement.* Oxford, UK: Berghahn Books.
McGahan, Hilly (2008, June 8). Guatemala dam will bring money, misery. San Francisco Chronicle.
Minority Rights Group International. (2008a). World directory of minorities and indigenous peoples—Guatemala. Accessed April 1, 2009. http://minorityrights.org/country/guatemala/
Montejo, V. (2005). *Maya intellectual renaissance: Identity, representation, and leadership.* Austin, TX: University of Texas Press.
Munck, R. (2005). *Globalization and social exclusion: A transformationalist perspective.* Sterling, VA: Kumarian Press.
Nolin, C. (2006). *Transnational ruptures: Gender and forced migration.* Surrey, UK: Ashgate Publishing Ltd.
O'Brien, T. (2007). *Making the Americas: The United States and Latin America from the age of revolutions to the era of globalization.* Albuquerque, NM: University of New Mexico Press.
Oxfam America. (2005). *Rodolfo Pocop: An Indigenous perspective of mining in Guatemala.* Oxfam America. Accessed April 15, 2009 http://www.oxfamamerica.org/articles/rodolfo-pocop-an-indigenous-perspective-on-mining-in-guatemala
Peet, R., & Hartwick, E. R. (1999). *Theories of development.* New York, NY: Guilford Press.
Raxche' (Demetrío Rodriguez Guajan). (1996). Maya culture and the politics of development. In E. Fischer, & R. M Brown (Eds.), *Maya cultural activism in Guatemala* (pp. 74–88). Austin, TX: University of Texas Press.

Rights Action. (2006). Xalala Dam Project: Water privatization, "development" & repression! [Chixoy Dam Revisited!]. Rights Action—Action Alert. Accessed October 15 2008. http://www.rightsaction.org/urgent_com/Xala_Dam_111206.html

Rist, G. (2004). *The history of development: From western origin to global myth*. London, UK: Zed Books.

Scudder, T. (2005). *The future of large dams: Dealing with social, environmental, institutional and political costs*. London, UK: Earthscan.

Sidaway, J. D. (2007). Spaces of postdevelopment. *Progress in Human Geography, 31*(3), 345–361.

Simon, D. (2003). Dilemmas of development and the environment in a globalizing world: Theory, policy and praxis. *Progress in Development Studies, 3*(1), 3–41.

Stanley, J. (2004). *Development-induced displacement and resettlement*. FMO research guide, forced migration online; a world of information on human displacement. Accessed February 10, 2009 http://www.forcedmigration.org/guides/fmo022/

Steinberg, M. K., & Taylor, M. J. (2008). Guatemala's Altos de Cuchumatán: Landscape changes on the high frontier. *Mountain Research and Development, 28*(3–4), 255–262.

Taylor, M. J., Moran-Taylor, M. J., & Ruiz, D. R. (2006). Land, ethnic, and gender change: Transnational migration and its effects on Guatemalan lives and landscapes. *Geoforum, 37*, 41–61.

Viscidi, L. (2004). *A history of land in Guatemala: Conflict and hope for reform*. Americas Program, Interhemispheric Resource Center. Accessed April 15, 2008 www.americaspolicy.org

Warren, K. B. (1998). *Indigenous movements and their critics: Pan-Maya activism in Guatemala*. Princeton NJ: Princeton University Press.

Washington Office on Latin America (WOLA). (2003). *November rights and development bulletin*. Washington D.C: WOLA.

Watts, M. (2004). *Human rights, violence, and the oil complex*. Niger Delta, Economies of Violence, Working Paper No. 2. University of California Berkeley.

Windsor, J. E., & McVey, J. A. (2005). Annihilation of both place and sense of place: The experience of the Cheslatta T'En Canadian First Nation within the context of large-scale environmental projects. *Geographical Journal, 171*(2), 146–165.

Witness For Peace. (1996). *A people dammed: The impact of the World Bank Chixoy Hydroelectric Project in Guatemala*. Washington DC, USA: A Witness for Peace Publication.

World Bank. (2004). DR-CAFTA: Challenges and opportunities for Central America. Washington DC: World Bank.

World Bank. (2008). *Doing business 2008 Guatemala: Comparing regulation in 178 economies*. Washington DC: World Bank. Accessed May 10, 2009. http://blog.doingbusiness.org/.m/2008/11/guatemala-going.html

Wright, R. (2004). *A short history of progress*. Toronto, ON: House of Anansi Press.

Ziai, A. (2007). *Exploring post-development: Theory and practice, problems and perspectives*. New York, NY: Routledge.

Chapter 4
Guatemalan Historical Context

Abstract It would be difficult to appreciate the events at Chixoy and Río Negro without a general understanding of Guatemalan history, with particular attention to the five centuries of conflict between Spanish colonizers and the Indigenous Maya peoples. This chapter provides historical context relevant to this study, while remaining within the general themes of development and Indigenous repression. Historical and contemporary topics relevant to Guatemala will narrow in scope in the following chapter, where I provide an overview of the events at Río Negro and the Chixoy Dam.

Keywords Guatemala history · Colonialism · Neoliberalism · Mayan peoples · Development · Violence

Along with much of what we now refer to as 'Latin America,' from the U.S.–Mexico border down to the tip of Chile at Tierra del Fuego, present-day Guatemala is a land of conquest, where much of its native constituents—the Maya and poor Ladinos, as well as the Xinca and Garifuna peoples—continue to endure the lingering legacy of colonialism (Chasteen 2006; Lovell 1988; Jonas 1991; Perera 1993; Handy 1984; Nelson 2009). Despite the ability for some groups to defend their lands and cultural practices (Lovell 1988; Montejo 2005), the lives and livelihoods of all Indigenous Mayas were forever changed upon the arrival of Pedro de Alvarado and his troop of Conquistadores in 1524. While colonial rule varied widely throughout Guatemala, the Spanish colonists, and later the Ladinos (people of mixed Spanish and Indigenous ancestry), dominated the Maya peoples by displacing them from their original lands, forcing them into serf labor on haciendas or agricultural plantations, and resettling them into locations where they could be more easily controlled. As interpreted by historical geographer Lovell (1988, 30).

> Spanish hegemony put Maya communities under immediate pressure to conform to imperial designs. A fundamental element of the Hispanic quest for empire was to organize space and to control population movement by founding towns and villages. Under the policy of congregación, which began in the highlands in the mid-sixteenth century,

N. Einbinder, *Dams, Displacement and Development*, SpringerBriefs in Latin American Studies, DOI 10.1007/978-3-319-51511-3_4

> thousands of native families were coerced from their homes in the mountains into new settlements (congregaciones) built around churches located, wherever possible, in open valley floors. For the Spaniards, congregación promoted more effective civil administration, facilitated the conversion of Indians to Christianity, and created centralized pools of labor that could be drawn upon in myriad ways to meet imperial objectives.

It is estimated that on the eve of Spanish arrival, up to two million Maya peoples lived in the mountains and tropical lowlands of what we now call Guatemala (Lovell and Lutz 1996, 399). Within a century, their numbers would be at about one-twentieth of that. Although warfare is partly to blame for the rapid depopulation, disease was the major cause of death, as was the subsequent effects from forced labor and seasonal migration (Lovell and Lutz 1996, 403).

Shortly after independence from the Spanish Crown in 1821, Maya peoples were subjected to the next round of conquest.[1] Through newly enacted liberal reforms and policies under Caudillos[2] such as Justo Rufino Barrios, Guatemala projected itself into the global market and, as a result, large sections of the country were further carved into monocultures of commodity crops, such as coffee, sugar, bananas, and cotton. Handy (1984) informs us that Indigenous labor was integral to the formation of these systems of large-scale production and export. Eduardo Galeano, in his eloquent history of the oppressed, *Open Veins of Latin America* (1973, 49–50), describes the situation of the Latin American peasant illustrative of the period after independence, and into the present day.

> They participate in an economic and social order which assigns them the role of victim—the most exploited of the exploited. They buy and sell a good part of the few things they consume and produce, at the mercy of powerful and voracious intermediaries who charge much and pay little; they are day laborers on plantations, the cheapest work force, and soldiers in the mountains; they spend their days toiling for the world market or fighting for their conquerors.

The production of coffee, which had success in establishing a middle class in neighboring Costa Rica, created further stratification in Guatemala between the landless majority and plantation owners (Chasteen 2006, 184). According to researcher Shankar (1999, 19), "the introduction of coffee, the 'great modernizer' can be seen as one of the major causes of past and present-day poverty, landlessness, increased monoculture, changing land tenure and settlement patterns, and seasonal migration." Lovell (1995, 138) also argues that "to the state coffee represented progress, civilization, and advancement; to Maya communities it meant loss of land and forced indentured labour… inequalities that began to emerge at the national level in the nineteenth century have yet to be redressed, especially in the countryside." As a result of the continuation of this system—where fertile terrain is expropriated for export production, or simply placed under control by the oligarchy

---

[1]Historical geographer Lovell (1988, 27), identifies three 'cycles' of conquest in Guatemala. These include: (1) That of the original Spanish intrusion; (2) The introduction to both internal and international capitalism; (3) The campaign of state terror.

[2]A political/military head, with authoritarian characteristics.

for future use—Guatemala remains to this day with the most unequal distribution of land in all of Central America, and one of the most extreme in the world, with no more than two percent holding roughly 75% of agriculturally productive land (Saldivar Tanaka and Wittman 2003, 6).

The turn of the twentieth century brought forth two significant positivist-inspired dictatorships, those of Manuel Estrada Cabrera (1898–1930), and General Jorge Ubico (1930–1944) (Handy 1984). Under Cabrera, the United Fruit Company (UFCo), a U.S.-based corporation later known as Chiquita, was granted 40% of the country as land base for its banana growing operations. In addition to land, the company was able to monopolize the railroad and electrical systems in the country (Jonas 1991). According to historian Handy (1984, 85).

> The UFCo, Minor Keith and Samuel Zemurray [owners] used their powerful influence on Guatemalan politics freely, always watchful to maintain the company's dominant position in the economy and intent on wringing whatever concessions they could from a weakened congress and accommodating dictators. UFCo's intervention in politics was, in the words of Guatemalan academic and politician Luis Cordoza y Aragon, "constant, open and bloody".

Cabrera's successor, General Jorge Ubico, a notorious authoritarian, made it legal for hacienda owners to kill rebellious workers, and reacted mercilessly to any so-called communist activity (Marcelo 2006, 15). Grandin (2004, 48–49) informs us that Ubico, through ruthless military coercion, secured cheap labor for the plantations and public works projects. In recalling the effects of the Ubico regime on his Maya-Achí ancestors in the remote valleys of Baja Verapaz, Osorio (2003, 2) states how "the government used the army to force the rural peasants to construct roads and to put telegraph wires throughout the country. The infrastructure benefited the rich… they [the Mayas] worked, yet did not receive salaries or food."

In 1945, the country voted in their first democratically elected president, Juan José Arévalo, after a 1944 coup orchestrated by students, military officials, and liberals (Schlesinger and Kinzer 1983, 30–35). A number of social reforms were created under his presidency, such as the legalization of labor unions, initiatives for gender and racial equality, and the foundation for health care and social security. President Jacobo Arbenz Guzmán followed Arevalo in 1951, yet ultimately failed in his bold attempt to permanently restructure the feudal economy and unequal social structures in his country (Perera 1993; Grandin 2004). Latin American historian Chasteen (2006, 260) remarks on Arbenz's brief era of revolution.

> In this country where half the people were illiterate Mayan peasants, treated more or less like animals by the owners of the coffee plantations, who retained great influence, Arbenz started to confiscate large estates, and divide them up for peasant cultivators. In addition, his government expropriated land from United Fruit, along with Guatemala's foreign-owned railway.

Through Arbenz's ambitious agrarian reform program, which benefited up to 100,000 families (Galeano 1973, 113), the state also began to ascertain control over United Fruit's monopoly by charging them taxes on their land and exports. It was not long before Washington decided that Arbenz was creating an unfavorable environment for business, especially that of United Fruit, which had its CIA connections (Schlesinger and Kinzer 1983). According to the powerful United Fruit elite among the Eisenhower administration, Arbenz and his administration had become a 'communist' threat—yet critics will argue that Arbenz represented simply a barrier to their financial interests, which had been protected in the past by dictators such as Ubico and Cabrera (Whitfield 1996, 158). It was not long after Arbenz's enactment of land reform that a U.S. orchestrated campaign for removal was underway (LaFaber 1993, 121). In a short yet very symbolic battle, Arbenz was overthrown in 1954 by the U.S. military and conservative Guatemalan partners, thus initiating the first of the cold war crusades in the Global South, and the extermination of other nationalistic movements of the period (Chasteen 2006, 293; Grandin 2007).

Colonel Castillo Armas took the presidential office and quickly reversed reforms from the previous ten years, including the abolition of all labor organizations (Perrera 1993). According to historian Handy (1984, 188), the movement towards 'ultra-laissez faire' economic development (increased export agriculture, large infrastructure projects, lower taxes), designed and assisted almost exclusively by the U.S. and the World Bank, resulted in deep inequalities—and social tensions—that still resonate today (Manz 2005; Nelson 2009). The switch in power and ideologies also left peasants without protection from the large landowners, who were recruiting the necessary military connections for increased power over workers (Jonas 1991). The Armas regime named supporters of the policies of Arbenz as insubordinate, or communists, and developed lists of the possible suspects (Booth and Walker 1993, 117). After six years of increased repression and militarization, a 1960 rebellion sparked by a group of dissident junior military officers initiated the 36-year internal armed conflict.

The history of the Guatemalan civil uprising—and the state-sponsored terror that followed in its wake—is complex, and the intensity of guerrilla and counterinsurgency activities varied throughout the country from one region to the next. Like much of Central America, crises relating to the firmly entrenched export driven economy and national debt, authoritarian repression by the military and paramilitary, and a growing socialist movement, caused simultaneous rebellions throughout the region (Fagan 1987, 22). As expressed in historian LaFeber's *Inevitable Revolutions* (1993), military training and assistance offered by the U.S., and particularly the CIA, were essential for the success of the Guatemalan army and their highly organized campaign against any leftist, socially progressive, or subversive movements that might challenge the free-market system. As pointed out by Manz (2005, 21):

> The growing political realities of the cold war and the United States foreign policy toward Central America gave the Guatemalan military and economic elites complete license to rule in an increasingly authoritarian way. Anticommunism served to justify and conceal the most heinous of crimes, and the United States—except for the Carter administration—eagerly funneled millions of dollars to military regimes decade after decade, showing no concern for the brutality committed by the armed forces. The battle for global, ideological hegemony had far-reaching local consequences for even the most isolated of peasants.

Although there is no single definitive motive for the uprising in Guatemala, Handy (1984) suggests its roots lie in the aborted revolution and agrarian reform under President Arbenz. According to Jonas (1991, 6), "The 'rebels' were not simply those who have taken up arms but are the unseen hundreds of thousands among Guatemala's 87% majority who have refused to accept a fate of poverty and discrimination." Along with the hundreds, if not thousands, of rebellions that have taken place in Guatemala, the poor distribution of land, coupled by social exclusion and a deeply entrenched structure of racism and inequality, must be acknowledged in order to fully understand the underlying causes of the armed conflict (Grandin 2004; Stewart 2006).

According to Lovell (1988, 44), the radicalization of the countryside, particularly in the early to mid 1970s, was aided by foreign-born Catholic Priests associated with Acción Católica (Catholic Action), a group which quickly "engaged in projects sought to ameliorate the social and economic life their native parishioners." Projects relating to the formation of agricultural cooperatives were particularly effective as a means of fomenting community independence and empowerment, thus producing a direct challenge to the Guatemalan status quo (Lovell 1988, 44), particularly that of plantation-based agribusiness, which at that time experienced a boom in export production and sales (see Davis 1988).

During the period of construction of the Chixoy Hydroelectric Dam, from 1976 to 1983—without a doubt the worst period of the entire conflict (see Carmack 1988; CEH 1999; Lovell 1988)—much of the nation experienced an unprecedented level of violence, described by Latin American scholar Jonas (1991, 146) as a "wholesale slaughter and genocide by the new death squads, the counterinsurgent forces, this time carrying out illegal violence themselves, without the façade of legal constraints." The combination of economic crisis, a widening gap between wealth distribution and the lowering of wages (despite marked growth in GDP), a continuation of unequal power structures and racism, and the earthquake in 1976, which left one million homeless and 23,000 dead, reinvigorated popular movements that originated in the early 1960s (Jonas 1991; CEH 1999; Booth et. al 2009, 120–121; Grandin 2004). As pointed out by anthropologist Davis (1988), the increase in landlessness of the Indigenous majority and the surge in economic productivity by the oligarchy—resulting in a strong push for cheap seasonal labor—increased the already explosive tensions between the two groups. Resistance movements—often with Marxist affiliation—in both the countryside and the cities were able to draw popular support from a desperate majority, with an increase in state-sponsored violence to follow (May 2001). According to Jonas (1991, 88):

> By the late 1970's the bourgeoisie had consolidated significantly with each other, and with the military, forming a tight alliance. The centuries of racism and exclusion, necessary for the elite to continue their financial success, had reached a boiling point.

Again, the demographic with the highest rate of casualty and victimization was the Indigenous peasant (CEH 1999), and at the height of the violence, just being Maya put one in a precarious, if not dangerous, position (Lovell 1988, 46).

Perhaps no other incident embodies this new epoch of violence more appropriately than the massacre at Panzos, Alta Verapaz on May 29, 1978, an event that Lovell (1988, 45) refers to as the likely beginning of widespread state terror. According to historian Grandin (2004), the Indigenous protest that took place that morning in the town square, and the shooting that followed, was not unlike so many others in the rural district of Alta Verapaz. However, as stated by Grandin (2004, 2):

> …the Panzos massacre is distinct in that it represents the passing of such exhausted patterns of protest and reaction, prefiguring more deadly forms of counterinsurgent violence that were soon to come…Likewise, the soldiers who guarded the plaza that Monday morning may have been called in by local planters, but they were no sleepy outland militia detachment occasionally roused into action. They were part of Guatemala's new army, steeped in anti-communism and flushed with counterinsurgent training equipment, the front line in an escalating civil war between a spreading rural insurgency and an increasingly repressive state.

More than one hundred Q'eqchi' Maya were killed on that day, and military units stayed in the region for one month, patrolling the area by land and air (Grandin 2004, 164).

The back-to-back military regimes of General Roméo Lucas García (1978–82) and General Efraín Ríos Montt (1982–83) followed this event, and with their anti-communist zeal and brutal counterinsurgency tactics, the massacres, executions, and disappearances escalated into genocidal proportions, especially in rural districts (Grandin 2004, 3; Jonas 1991, 122). Under these two administrations, which enjoyed personal and financial support from U.S. President Ronald Reagan (Manz 2005, 23; Fagan 1987; Perera 1993), a scorched earth policy was adopted, which involved the burning of whole villages containing suspected guerrillas or guerrilla sympathizers (CEH 1999). According to the UN Truth Commission (CEH 1999) at least 626 primarily Maya occupied villages were completely annihilated through this policy. The military also set up armed civilian defense patrols (Patrullas de Autodefensa Civil—PAC) (Colajacomo 1999; Remijnse 2001) as a means of controlling the insurgency in rural districts. Up to one million men (mostly Maya) were eventually forced into the PACs, effectively acting as an extension of the army's fear campaign throughout the rural areas (Jonas 1991, 150; Remijnse 2001).

Another element of Ríos Montt's strategic control over the rural areas was the creation of 'model villages,' or, 'development poles,' which were in essence camps built to house and monitor internal refugees. As Guatemalan writer Perera (1993, 111) states:

> For all their frequently touted resemblance to the Vietnam War's notorious "strategic hamlets," the model villages more accurately reflect the colonial Spaniards' reduction of rebellious Indian communities into nucleated settlements for the convenience of landowners and the missionary priests.

In reflecting on the short yet highly destructive presidency of Ríos Montt, historian Handy (1984, 255) remarks that "throughout 1982 and 1983 the military was able to complete a process that began in 1542: the acquisition of control over the countryside and the total breakdown of village autonomy."

Following a military coup that ousted Ríos Montt from presidential power, a drawn-out movement towards peace was initiated with the democratic December 1985 election of civilian president Vinicio Cerezo Arévalo (Lovell 1995). Despite Cerezo's attempt to initiate dialogue with the URNG[3] (United Revolutionary Unity of Guatemala), the military, at least at this stage, refused and continued its unlawful detentions and assassinations in the name of crushing the dying insurgency (Jonas 2000). It was not until 1996, with the facilitation of the United Nations, that military and guerrilla groups signed a final Peace Agreement (Jonas 2000).

In all, the U.N. sponsored Truth Commission estimates that the military and paramilitary death squads killed at least 200,000 thousand people during the conflict, with over one million persons displaced (CEH 1999). Although much of the killings and massacres were aimed at the rural, Indigenous populations, as the military viewed Maya peasants as the most susceptible to be taken in by the guerrilla movement (May 1999, 73), political assassinations of possible left wing conspirators (teachers, unionists, journalists, intellectuals, students) were all too common in the capital as a means of social cleansing (Galeano 1973; Hanlon and Shankar 2000), with a tally of over 34,000 Ladino deaths (Taylor 2005).

Despite the high hopes that accompanied the signing of the Peace Accords in 1996, disparities have not improved. According to the UNDP (see CESR and ICEFI 2008, 2), Guatemala has the highest income inequality in Latin America. Poverty stands at over 80%, and despite continued talk of reform, the country still suffers from one of the most unequal land distribution patterns in the world, with the vast majority of fertile land is owned and occupied by large export-agricultural plantations (Godoy 2006; Hamilton and Fischer 2005).

On top of these economic realities, 'post-conflict' Guatemala still reels in its legacy of violence and insecurity (Benson et al. 2008; Renwick 2016; Smith and Offit 2010). In a reflection of life after 1996, anthropologists Benson et al. (2008, 29) argue that despite what the indicators show—that the frequency of political murders and disappearances have dissipated, and that massacres no longer occur in the rural districts of the county—"Peace has proven elusive."

What they mean by this, as I witnessed on my own visits to the country, is that violence and insecurity still prevails in present-day Guatemala, just in a different form. At the time of my research, gang violence and organized crime, primarily in the capital, had reached unprecedented levels, with the average murder rate at 17 per day (GHRC 2009). Extortion, femicide, narco-related violence, and assassinations

[3]The Unidad Revolucionaria Nacional Guatemalteca (URNG) was formed as an umbrella for the four main guerrilla groups operating within the country. These groups include: The Guerrilla Army of the Poor (EGP), Rebel Armed Forces (FAR), Revolutionary Organization of the People in Arms (ORPA), and the National Directing Nucleus of PGT (PGT-NDN). The URNG was established as a legitimate political party in 1998.

have risen every year since 2000 (Sanford 2008). Due to the low incidence of investigation by the national police—which remains currently at two percent—impunity reigns, and a resurgence of 'mano dura' (strong arm) tactics offered by right wing political groups—often associated with former military regimes of 80s and 90s—seems a likely outcome in the future (Melville 2007; GHRC 2007).

The source of this new violence appears to be another flare up from unresolved social and economic problems that afflict the country, as well as a destabilization created by the oligarchy/military to ensure that power remains consolidated, and unchallenged by renewed popular social movements (Campbell 2007; Isaacs and Schwartz 2013). As stated by geographer Manz (2008, 151):

> The violence is rooted in a society that historically has been deeply divided along ethnic and rigid class lines and that has been fundamentally unjust, discriminatory, and abusive toward the oppressed population. The challenge then is much greater, and the likelihood of continued violence that much more likely, when the underlying grievances have not been and are unlikely to be addressed because those in power do not have the will to bring about the necessary changes.

Neoliberal restructuring of the Guatemalan economy over the past two decades, which has resulted in a decrease in social services and regulation for corporations (particularly in the mining sector), as well as an increase in mega-project developments that further encroach upon lands occupied by Indigenous campesinos (Holden and Jacobson 2008), is also noted as a probable underlying cause in Guatemala's overall increase in violence and insecurity among its populace (Benson et al. 2008). As stated by sociologist Robinson (2000, 1), Guatemala's neoliberal order "is inherently unstable and indicates contradictions internal to global capitalism, including social polarization between the rich and the poor, the loss of nation-state autonomy and regulatory power, and the deteriorization of the social fabric in civil society." Kurtenbach (2008), a principal researcher at the Inter-American Dialogue in Washington D.C, claims that despite macroeconomic growth over the past decade, little can be said regarding improvement in terms of social inclusion and overall equality, which has lead to the increase in crime. Kurtenback (2008, 35) also argues that through these neoliberal reforms, most outside capital is directed towards mega-projects, "in mining or energy sectors which only provide few job opportunities and where the profit is made by a small group of local or international entrepreneurs…. Thus, natural resources are not used for the public good but for private enrichment, another process that can be interpreted as path dependency."

Mining, as well as the development and planning of other new mega-projects, is causing considerable social unrest in certain departments, as the newly ratified mining law makes it almost impossible for local Indigenous occupants to legally protect their land (Imai et. al 2007; America 2009). According to Guatemala's Ministry of Energy and Mines, there were 356 mining licenses granted as of December 2006, with hundreds more in the process.

While resistance to this kind of imposed 'development' has increased dramatically over the past several years (International Rivers 2008; McGahan 2008;

Paley 2008)—as outlined in the previous chapter—there has yet to be a conclusive agreement between communities opposed to the presence of mining in their districts, and those who grant licenses to the foreign corporations.

Although this chapter presents just a slice of historical information pertaining to the lands and peoples of Guatemala, I chose very specific incidents and facts regarding conflict over territory and resources, which will allow for a better understanding of the Chixoy Dam and its effects on the peoples of Río Negro. In the following chapter, I expand on the region in which Río Negro is situated and the events that regard the Chixoy Dam, with specific reference to the larger picture covered in this chapter.

References

America, O. (2009). Metals mining and sustainable development in Central America: An assessment of benefits and costs. Oxfam America. Accessed 1 June 2009. http://www.oxfamamerica.org/publications/metals-mining-and-sustainable-development-in-central-america

Benson, P., Fischer, E. F., & Thomas, K. (2008). Resocializing suffering: Neoliberalism, accusation, and the sociopolitical context of Guatemala's new violence. *Latin American Perspectives, 35*(5), 35–58.

Booth, J. A., Wade, C. J., & Walker, T. (2009). *Understanding Central America: Global forces, rebellion, and change*. Oxford, UK: Westview Press.

Booth, J. A., & Walker, T. W. (1993). *Understanding Central America*. Oxford, UK: Westview Press.

Campbell, M. (2007). *The shadow of war*. London, UK: Amnesty International Magazine.

Carmack, R. M. (Ed.). (1988). *Harvest of violence: The Maya Indians and the Guatemala crisis*. Norman, Oklahoma: University of Oklahoma Press.

Centre for Economic and Social Rights (CESR) and Instituto Centroamericano de Estudios Fiscales (ICEFI). (2008). Economic, social and cultural rights in Guatemala: A selective briefing. Submitted to the Office of the High Commissioner for Human Rights on the occasion of the second session of the Universal Periodic Review, May 2008. www.cesr.org/.../UPR%20Guatemala%20submission%202008.pdf

Chasteen, J. C. (2006). *Born in blood & fire: A concise history of Latin America*. New York, NY: W.W. Norton and Company.

Colajacomo, J. (1999). The Chixoy Dam: The Maya Achi' genocide. The story of forced resettlement. Contributing paper, World Commission on Dams. Cape Town, South Africa. Accessed 1 March 2008. http://www.dams.org/kbase/thematic/tr12.htm

Commission For Historical Clarification (CEH). (1999). Guatemala: Memory of silence. Report of the Commission for Historical Clarification Conclusions and Recommendations Guatemala. http://hrdata.aaas.org/ceh/report/english/

Davis, S. H. (1988). Introduction: Sowing the seeds of violence. In R. M. Carmack (Ed.), *Harvest of violence: The Maya Indians and the Guatemalan crisis*. Norman, Ok: University of Oklahoma Press.

Fagan, R. R. (1987). *Forging peace: The challenge of Central America*. New York, NY: Basil Blackwell.

Galeano, E. (1973). *Open veins of Latin America*. New York, NY: Monthly.

GHRC (Guatemala Human Rights Commission). (2005). Indigenous and land rights leaders threatened. Accessed 5 January 2009. http://www.ghrc-usa.org/Resources/2005/IndigenousLeadersThreatened.htm

Godoy, A. S. (2006). *Popular injustice: Violence, community, and law in Latin America*. Stanford, CA: Stanford University Press.
Grandin, G. (2004). *The last colonial massacre: Latin America in the Cold War*. Chicago, IL: University of Chicago Press.
Hamilton, S., & Fischer, E. F. (2005). Maya farmers and export agriculture in Highland Guatemala: Implications for development and labor relations. *Latin American Perspectives, 32* (33), 33–58.
Handy, J. (1984). *Gift of the devil*. Toronto, ON: Between The Lines.
Hanlon, C. N., & Shankar, F. (2000). Gendered spaces of terror and assault: The testimonio of REMHI and the Commission for Historical Clarification in Guatemala. *Gender, Place and Culture*, *7*(3), 265–286.
Holden, W. N., & Jacobson, R. D. (2008). Civil society opposition to nonferrous metals mining in Guatemala. *Voluntas*, *19*, 325–350.
Imai, S., Mehranvar, L., & Sander, J. (2007). Breaching indigenous law: Canadian mining in Guatemala. *Indigenous Law Journal, 6*(1), 102–129.
International Rivers. (2008). Financial crisis stops Xalala Dam—For now. International rivers, Berkeley California. Accessed 15 June 2009. http://www.internationalrivers.org/node/3538
Isaacs, A. & Schwartz, R. (2013). Repression, resistance, and indigenous rights in Guatemala. *Americas Quarterly*. Accessed 14 November 2016. http://www.americasquarterly.org/content/repression-resistance-and-indigenous-rights-guatemala
Jonas, S. (1991). *The battle for Guatemala: Rebels, death squads and U.S. power*. New York, NY: Westview Press.
Jonas, S. (2000). *Of centaurs and doves: Guatemala's peace process*. New York, NY: Westview Press.
Kurtenbach, S. (2008). Guatemala's post-war development: The structural failure of low intensity peace. Project working paper no. 3. Institute for Development and Peace (INEF). Germany: University of Duisburg-Essen.
LaFaber, W. (1993). *Inevitable revolutions: The United States in Central America*. New York, NY: W.W. Norton & Company.
Lovell, W. G. (1988). Surviving conquest: The Maya of Guatemala in historical perspective. *Latin American Research Review, 23*(2), 25–57.
Lovell, W. G. (1995). *A beauty that hurts: Life and death in Guatemala*. Austin, TX: University of Texas Press.
Lovell, W. G., & Lutz, C. H. (1996). "A Dark Obverse": Maya survival in Guatemala 1520–1994. *The Geographical Review, 85,* 398–407.
Manz, B. (2005). *Paradise in ashes: A Guatemalan journey of courage, terror, and hope*. Los Angeles, CA: University of California Press.
Manz, B. (2008). The continuum of violence in post-war Guatemala. In A. Waterston (Ed.), *The Anthropology of war: Views from the frontline*. Oxford, UK: Berghahn Books.
Marcelo, B. (2006). Good dictator, bad dictator: United Fruit Company and economic nationalism in Central America in the Twentieth Century. Retrieved 22 Jan 2008, from http://www.buisiness.uiuc.edu/Working_Papers/papers/06-0115.pdf
May, R. (2001). Terror in the countryside: Campesino responses to political violence in Guatemala, 1954–1985. Center for International Studies, Ohio University, U.S.A.
McGahan, H. (2008, June 8). Guatemala dam will bring money, misery. *San Francisco Chronicle*.
Melville, K. (2007). Confronting femicide in Guatemala. *Upside Down World*. http://upsidedownworld.org/main/content/view/879/33/
Montejo, V. D. (2005). *Maya intellectual renaissance: Identity, representation, and leadership*. Austin, TX: University of Texas Press.
Nelson, D. M. (2009). *Reckoning: The ends of war in Guatemala*. Durham, NC: Duke University Press.
Osorio, J. (2003). *The Río Negro massacres*. Toronto, ON: Jesús Tecú Osorio & Rights Action.
Paley, D. (2008, June 21). Goldcorp: Occupation and resistance in Guatemala. The Dominion. http://www.dominionpaper.ca/weblogs/dawn/1887

Perera, V. (1993). *Unfinished conquest*. Los Angeles, CA: University of California Press.

Remijnse, S. (2001). Remembering civil patrols in Joyabaj, Guatemala. *Bulletin of Latin American Research, 20*(4), 254–269.

Renwick, D. (2016). Central America's violent Northern Triangle. Council of Foreign Affairs. Accessed 14 November 2016. http://www.cfr.org/transnational-crime/central-americas-violent-northern-triangle/p37286

Robinson, W. I. (2000). Neoliberalism, the global elite, and the Guatemalan transition: A critical macrosocial analysis. *Journal of Interamerican Studies and World Affairs, 42*(4), 89–107.

Saldivar Tanaka, L., & Wittman, H. (2003). The agrarian question in Guatemala. Land Action Research Network (LRAN). Accessed 15 February 2009. http://www.landaction.org/display.php?article=54

Sanford, V. (2008). From genocide to femicide: Impunity and human rights in twenty-first century Guatemala. *Journal of Human Rights, 7*, 104–122.

Schlesinger, S., & Kinzer, S. (1983). *Bitter fruit: The untold story of the American coup in Guatemala*. Garden City, NY: Anchor Books.

Shankar, F. (1999). Organic beans and ethical aromas: Small-scale coffee production in Guatemala (Unpublished master's thesis). Department of Geography, Queen's University, Kingston, Ontario.

Smith, T. J. & Offit, T. A. (2010). Confronting violence in postwar Guatemala: An introduction. The Journal of Latin American and Caribbean Anthropology, 15(1).

Stewart, J. (2006). When local troubles become transnational: The transformation of a Guatemalan indigenous rights movement. In H. Johnston & P. Alemida (Eds.), *Latin American social movements: Globalization, democratization, and transnational networks* (pp. 197–214). Maryland, USA: Rowman and Littlefield.

Taylor, M. J. (2005). Electrifying rural Guatemala: Central policy and rural reality. *Environment and Planning C, 23*, 173–189.

Whitfield, S. J. (1996). *The culture of the Cold War*. Baltimore, MD: John Hopkins Press.

Chapter 5
Río Negro and the Chixoy Dam

Abstract This chapter provides an overview of the Chixoy Dam hydroelectric project and its effects on the Maya-Achí population from Río Negro, as well as a brief description of the regional and human history. Information presented in this section was drawn from a variety of sources, including publications from NGO's, journalists, and academics, as well as evidence from ethnographic research carried out in the region during the winter of 2009. A more in-depth look into the specific consequences of this project on individuals as well as the Río Negro community as a whole will be explored at length in the following analysis and discussion chapters.

Keywords Hydroelectric dams · Mayan peoples · Massacres · Civil conflict · Genocide · Development · Neoliberalism · Violence

The Chixoy River, or Negro, as it is also referred to, is one of Guatemala's largest rivers, originating in the moist western slopes of the Cuchumatanes Mountains in the departments of El Quiché and Huehuetenango. After making a large S-shaped bend through the deep canyons of north-central Guatemala, the Chixoy flows due north through the lowlands of the Ixcán region. Once reaching Mexico it becomes the Río Usumcinta, and after a long winding stretch through Chiapas and Tabasco arrives to the Bay of Campeche at the Caribbean Sea.

At the river's bend approximately 150 km north of Guatemala City lies the remote and relatively low (700 m elevation) fertile basin once referred to as *La Cuenca Media del Río Chixoy*, or, the Middle Basin of the River Chixoy (Ichon 1978). Through archeological research, we know that various Indigenous Maya groups have inhabited the region since the Classic Period (300 BC to 900 AD) (Colajacomo 1999, 3), with dozens of religious sites (now mostly underwater) marking the existence of an extensive network of communities and places of worship. One site in particular, the Great City of Cauinal (see Rodriguez 2009), a sprawling ruin emerging from the south end of the Chixoy Reservoir at the end of each dry season, has been compared in importance to that of the Maya archeological centers found on the Yucatán Peninsula (Arnauld 1996).

N. Einbinder, *Dams, Displacement and Development*, SpringerBriefs in Latin American Studies, DOI 10.1007/978-3-319-51511-3_5

Before 1982, some 3450 people from 23 communities lived in the region, utilizing the rich farmland surrounding the river and local fishery, and contributing to a vibrant local economy (Colajocomo 1999; Johnston 2005a). Local centers of trade, such as Los Encuentros, situated at the confluence of the Chixoy and Salama rivers, were established for residents to sell their goods, as well as celebrate Catholic and Maya holidays (Johnston 2005a). The community of Río Negro, or Panima' in the Achí language, was one of the larger villages along this stretch of the river, with a total of 791 inhabitants according to a 1977 census (Colajocomo 1999, 1). According to previous residents the original inhabitants arrived from Rabinal no more than a couple hundred years before. Reasons for the migration included that of access to good land for their animals, cornfields, and wood collection, as well as a dependable source of clean water originating from springs above the valley.

Like most rural Guatemalan communities, inhabitants from Río Negro were primarily subsistence farmers, with small family plots, or *milpas*,[1] which extended up the hillsides above the valley floor. According to SEBASTIAN, who lived there until 1982, numerous species of fruit trees lined the river—mangos, jocote, banana, papaya, peanuts, and citrus, among other native varieties. Other past inhabitants recall the bountiful harvests of corn and fruit, as well as access to excellent fishing. The community had a school, church, and a local development committee (Witness For Peace 1996). On market days in Rabinal, six to eight hours away, vendors from Río Negro would leave before dawn, on foot, to sell excess goods and crafts, and return with items they could not grow or make for themselves (Osorio 2003). Residents were also known for their work with local palm fronds (and still are, to some extent), which they collected locally and wove into mats, called *petates*, and sold at various markets.

Without roads or profitable land for agro-industrial development the region was isolated, and insulated; and for the most part, able to exist without the intrusion of colonial policies that dominated elsewhere (see Lovell 1988). French archeologist Ichon (1978, 18–21), who surveyed the basin for Maya ruins, wrote extensively about the parallels he observed at Río Negro and what he knew about pre-Hispanic life among the Maya. Instead of the typical Spanish colonial layout, with a town square or plaza surrounded by homes, churches, and other community structures, Río Negro consisted of a scattering of family ranchos, or compounds, surrounded by individual *milpas*, and other food producing trees. Ichon (1978) estimated that about half of the families had cattle, but that raising chickens and pigs was more common. He also noted the strong manufacturing co-ops for *petates* and other crafts.

Former residents of Río Negro speak of a difficult life, but one that was provided for, and tranquil (Johnston 2005a; Witness For Peace 1996). As stated by CARLOS, a local leader who lived in Río Negro during his childhood and into his twenties

[1]Endemic to Mesoamerica, *milpa* farming is an intercropped system perfected over the course of thousands of years. The primary crop is maize, typically a local variety, along with squash, beans, and often a host of other native, edible herbs (see Mann 2006).

> Before in Río Negro, we made *petates* and we fished. This was the basis of our economy. Similarly, one might have cows, or pigs, or chickens. It was all free land: we had ample resources. Occasionally people would leave for the coast to cut sugarcane to earn money, during the month of July or January. But it was for just one month per year, so we could buy clothes and medicine.

5.1 The Chixoy Dam Project

As stated in the previous chapter on Guatemala history, the 1970s and 1980s—a period also known as *La Violencia*—were terrible, unpredictable times, even in the most remote districts of the country (see Carmack 1988). The municipality of Rabinal, which included the community of Río Negro, experienced an unusually high level of military repression, resulting in the murder and disappearance of roughly 5000, principally Achí, civilians (CIIDH 1999, Chap. 4). According to sociologist Julie Stewart (2006, 203), Rabinal is "a microcosm of Guatemala." With its high levels of poverty, as well as its historic land struggles between Ladinos and the Maya-Achí majority, the municipality was able to draw early popular support from the FAR (Revolutionary Armed Forces), as well as the EGP (Guerrilla Army of the Poor) (CIIDH 1999, Chap. 4). As a result, the military government labeled many individuals, or entire communities, as guerrillas or guerrilla supporters, resulting in widespread massacres and disappearances (Stewart 2006).

Incidentally, *La Violencia* coincided with the global push for increased modernization and free market economic policies, particularly within developing nations of the Global South (Jonas 1991, 75; Booth and Walker 1993). As pointed out by O'Brien (2007, 266) the U.S. government, along with international banking firms, vigorously promoted this new 'religion' of policies and development—what we now refer to as neoliberalism—with subsequent aid flowing into the hands of right-wing military governments throughout Latin America. At the same time, Guatemala enjoyed a surge in export-driven production (Davis 1988); and as a result of a rapidly industrializing economy, as well as a spike in oil and gas prices, the country experienced a deepening energy crisis (Witness For Peace 1996; World Bank 1978).

Amidst rolling blackouts, Guatemala's National Institute of Electrification (INDE)[2] designed a plan for a hydroelectric facility in the Chixoy River basin, located at Pueblo Viejo, about six kilometers downstream from the village of Río Negro. Although project plans began as early as 1972 (Witness For Peace 1996,

[2]The INDE was founded under the administration of Manuel Ydígoras Fuentes (1957–63), as a means to gain military control over the nation's resources. An ex-General under Gorge Ubico (see previous section), Ydígoras was known for his "unparalleled corruption" and was overthrown from office after massive protests (Handy 1984, 190).

11), the idea was sold in 1976 to the World Bank with an initial loan of $72 million, and the Inter-American Development Bank for $105 million, with an estimated total cost of $340 million (Colajacomo 1999, 2; World Bank 1978, 49). According to the Witness for Peace report (1996, 11): "international lending institutions quickly approved funds for this 'engineering miracle' which was to transform Guatemala's 'third world' economy with cheap energy by the year 2000."[3] As stated by independent researcher Lynch (2007, 4), "[Chixoy] was intended to provide some 60% of Guatemala's electricity. It would be the energy source that would permit exploitation of copper and nickel deposits—and possibly oil—in the Northern Transverse Strip."[4]

It was not until funding was secured that INDE officials approached Río Negro inhabitants to notify them about the dam project and their need to relocate, due to the flooding that would occur behind the dam. After extensive research in the region, anthropologist Johnston (2005a, 16) concluded that no agreement or dialogue ever existed between INDE and the affected communities, and that there was no legal transfer of land titles. Early INDE documents indicate no plan for social impact assessments or resettlement, and that the agency only began to devise a plan based on World Bank conditions for the loan (Lynch 2007, 9). Former residents assert that INDE officials simply showed up by helicopter and told them that their village would be flooded, and that they would provide a *finca*, or farm, where they could move. According to local community leader Osorio (2005, 13), "They [INDE] promised to finance the construction of one hundred and fifty homes, purchase farming land and cattle pasture. Those affected would choose the land that would replace their flooded land. The INDE promised to compensate people who would lose their crops."

It was early in 1980 that INDE officials led a handful of community leaders across the mountains to the 'model village' of Pacux, near Rabinal, which was to be their new home. The restricted, urban setting, and lack of communal land to grow food and continue their subsistence lifestyle led to the conclusion that they must either fight to obtain proper compensation (of which they were entitled, and specifically funded for under IDB and World Bank policies), or resist leaving Río Negro (Witness for Peace 1996, 17).

According to the Witness for Peace (1996) report, as well as testimonies from survivors, varying degrees of resistance existed among the community, but most residents decided they wanted to stay in Río Negro and protest the move, especially after seeing Pacux, and realizing that the INDE were not going to fulfill their

---

[3]It should be noted that these loans were given to the state-run electricity company without any proof that they had title to the land needed for the reservoir or dam (Johnston 2005a, b, 3).

[4]Known as the 'zone of the generals' (Jonas 1991, 128; Manz 1981), this lowland region bordering Mexico was bought up in large quantities (100's of thousands of acres) by President General Lucas García, and other military generals of the era, for investments in natural resource extraction. Manz (1981), reports of a massive infrastructure program in the region—including hydroelectric power facilities—to help aid the efficiency of projects such as oil extraction, coffee production, and logging.

promises.[5] Ichon (1978, 27) indicates in his report that the elders of the community were most resistant, due to their strong attachment to the land. The archeologist also noted that reactions between the affected communities varied radically. For example, the strictly Indigenous population of Río Negro (rather than a mixed Ladino/Achí population, as was the case in other communities), as well as its strong social cohesion among residents, contrasted sharply with the neighboring village of Chicruz, whose relocation came with little resistance (Ichon 1978, 31).

The organized—yet peaceful, and completely legal—resistance of Río Negro citizens was met with a campaign of terror and intimidation by military officials, and later by the Civilian Patrol (PAC) from the neighboring community of Xococ, which had effectively aligned itself with right-wing military groups[6] (Witness for Peace 1996; Osorio 2003). Because of Río Negro's opposition to the construction of the dam and their forced displacement, INDE and the military labeled residents as subversive and in coordination with guerrillas (Johnston 2005a, 4).[7] According to survivors the violence began in 1979, soon after INDE officials demanded land titles from Río Negro residents, and assured that they would be returned. When village leaders demanded for their titles several months later, INDE claimed to have never received them. Following this incident, tensions mounted between INDE officers and local residents, and during one such heated argument, military police (contracted by INDE) shot and killed seven individuals from Río Negro (Lynch 2007, 13). One officer was chased away from the village and subsequently drowned in the river. Accusations by the military of guerrilla corroboration mounted against villagers, and officers began to visit the community on a regular basis to interrogate and harass potential suspects. Not long after this initial incident, two Río Negro community representatives were found mutilated after handing over to the INDE documents which outlined compensation promises, land title, and signatures (Lynch 2007, 14).

As noted by Osorio's (2003, 21–22), it was during the month of February 1982 when the Xococ PAC began to terrorize Río Negro residents, thus initiating a series of massacres that would result in over 400 deaths (CEH 1999, VI, Illustrative Case

[5]According to the exhaustive report published by Witness for Peace in 1996, the promises made by INDE included: (1) Cement houses; (2) Potable water and FREE electricity; (3) Five acres of fertile land per household; (4) A community truck; (5) Compensation for crops and orchards lost to flooding; (6) A church, schools, health center and access roads; (7) A boat; (8) Social services.

[6]The reasoning behind Xococ's affiliation with the military's extreme right-wing faction is complicated at best. Osorio (2003, 18–22) informs us that conflict between the military and EGP guerrillas (some with potential affiliation to Río Negro) resulted in assassinations and intimidation in the community (Xococ) during the early 1980s, and that through these complicated events, a PAC was formed to carry out insurgency operations under the military's close supervision and intimidation (see Stewart 2006).

[7]According to Osorio's (2003) testimony, as well as other respondents who participated in this study, the EGP guerrilla group became active in the region after dam construction began. Due to the fact that some residents in Río Negro worked for INDE in construction, as well as the diverse levels of resistance in the village, EGP presence caused tension, and made it easier for the government to label the community as a stronghold.

Annex no. 1). The first of these massacres took place in Xococ, after a local military commander ordered 72 men and women from Río Negro to report to the village on February 13 to retrieve identity cards which were confiscated a week prior. According to SEBASTIAN, as well as other supporting documents and testimonies, just one person escaped the massacre that took place, a 47–year-old woman named Theodora, who walked back to Río Negro and informed residents what had happened. Following this incident, soldiers appeared regularly and the older boys and men began to sleep out in the fields to avoid confrontation. One month after the Xococ massacre, on March 13, 1982, the PAC reappeared alongside military soldiers and captured roughly 200 women and children from their homes in Río Negro. After forcing them up the mountainside behind the village the captors raped, tortured, and eventually murdered 70 women, and 107 children at a mountain pass called Pocoxom, two kilometers from the river (Lynch 2007, 14; Witness for Peace 1996, 17). "The women didn't leave," one survivor stated. "Because no one thought that they would kill women, kids, or the elderly. We didn't think…" Through extensive work in the region the UN Truth Commission report (CEH 1999, VI, Illustrative Case Annex no. 1) describes the incidents from the March 13 massacre at Pocoxom, later interpreting it as genocide

> They then forced the people grouped together to walk three kilometers uphill. "During the entire walk, they hit the women a lot, they called them cows, they treated them as if they were cows rustled into a new field. They hit the children a lot and said they were children of the guerrilla." When arriving at the zenith of Pacoxom hill, a member of the Army, according to a declarant, said that "now it will be easy to kill some guerrillas." In this manner, they continued to torture and kill the passive victims. Some were hung from trees, others were killed by machete and others were shot. "The small children were held by their hair and hit against rocks to kill them." In one mass grave, they were throwing the cadavers. "One that was still alive suffering was left there like a log; one on top of another, they were not left in an orderly manner because they were just thrown in there." The grave was covered with rocks and branches. Around five o'clock in the afternoon the massacre was over and the assassins headed to Xococ. The aggressors took nineteen child survivors to Xococ.

The excuse for the massacres, as reported by the military, was that of a routine counterinsurgency operation (Colajacomo 1999, 5). A Río Negro survivor responds to this unquestionably false and ungrounded accusation

> How can innocent children and women, many of them pregnant, be mistaken for guerrillas? They couldn't be. None of us were. We were peasants trying to make a living from the soil like our parents and ancestors. I'll tell you the real reason for the violence: they wanted our land for their cursed reservoir and dam, and we were in the way (Witness For Peace 1996, 18).

Survivors of the massacre at Pocoxom fled to the steep mountains and canyons above the river valley, or were taken in as slaves by the paramilitaries, as was the case for Osorio (2003). According to survivors and documented in reports (Lynch 2007, 14; Witness For Peace 1996, 16), two more massacres succeeded that of March 13: one in May, where 82 persons were killed in the nearby village of Los

Encuentros, and then in September, where another 92 individuals were murdered near the Chixoy Dam at Aguas Frías.

It was in early 1983 that the valley was flooded from the newly completed dam, effectively inundating any structures that remained at Río Negro (Colajacomo 1999, 5). After hiding for nearly two years in the rugged mountains surrounding the basin, without proper shelter, food, or water, massacre survivors arrived to Pacux to live in the resettlement community built for them by the World Bank and the INDE.[8] Upon arrival some of the men, such as SEBASTIAN and other participants of this research, were locked in the adjacent military base for up to two weeks where they were tortured. After moving into the homes built by INDE, the military forced residents to create a PAC and continued to face constant harassment by army officials and other paramilitaries.

According to Osorio's (2003, 41) testimony, "The army entered the village of Pacux and entered the widow's homes and raped them. The patrollers did not object to the soldiers' activities because they knew they would accuse them of being guerrillas." Survivors from Río Negro continued to be kidnapped, tortured, and killed if they showed any signs of resistance against the conditions. In addition, INDE disregarded nearly all of their relocation promises. As stated in the Chixoy Legacy Issues Study (Johnston 2005a, 26):

> All the resettlement communities experience problems from the lack of potable water, which has caused disease; problems with crowded homes and deteriorated buildings; inadequate land to grow food for the household; and, lack of income generation opportunities. For more than 20 years these communities have suffered from the lack of access to fertile lands, markets, and critical resources (fish, palm leaves, fruits, firewood). Extreme poverty has contributed to malnutrition, and many died for lack of food in the first years of resettlement.

Initial reports by the World Bank make no reference to the mass killings that occurred at Río Negro, and another loan of $44 million was granted to the INDE after technical difficulties started in 1985 (Colajacomo 1999, 17). Although the role of individual Bank employees in the massacres is unclear, it is evident that the lending institutions knew about the violence (Witness For Peace 1996), as well as the fact that major difficulties involving resettlement and proper compensation existed (Johnston 2005a, 3). For example, Johnston (2005a, 17) found in her research that "World Bank and Inter-American Development Bank regularly sent staff to the area to evaluate performance and conduct new feasibility studies in support of additional financing." Through extensive research in the region, Stewart (2006, 206) indicates that many individuals living in the area knew of INDE's involvement in the massacres, stating that it was easier—and more cost efficient—to label those from Río Negro as guerrillas and eliminate them, rather than give them the compensation originally promised. Speculation also exists among local

[8]It should be noted that a handful of survivors made it to refugee camps in Alta Verapaz, Guatemala City, and in a rare circumstance through international adoption, the United States (see Flynn and McConahay 2002).

residents, survivors, and researchers concerning the involvement of INDE officials with corruption at the highest levels of the military (Stewart 2006; Witness for Peace 1996). As indicated by the Centre on Housing Rights and Evictions (COHRE) (2004, 14), the INDE was placed under military control by the administration of General Romeo Lucas García. Given the nature of this military government—and the investments the president, as well as other generals had in the Northern Transverse Strip (Jonas 1991, 128)—COHRE (2004, 36) documents conclude that: "INDE stood to benefit from the human rights violations suffered by the Río Negro community. The massacre, forced eviction and Río Negro community meant that INDE did not have to relocate or provide just and fair compensation to the residents, and as such INDE was unjustly enriched by these human rights violations."

The past 25 years have proved difficult for the survivors of the Río Negro massacres. Along with every other Chixoy-resettlement community, residents in Pacux suffer from extreme poverty as a result of a lack of economic opportunity and inadequate land in which to grow crops, as well as an increase in gang violence among its youth (COHRE 2004). As concluded from my research in the region during Winter 2009, as well as others (Colajacomo 1999; Johnston 2005a, b), it appears that despite the political action initiated by residents over the past 15 years in demanding reparations from the government, as well as justice for the murders that took place, little has transpired.[9]

As will be discussed in greater detail in the following chapter, as well as in the final conclusions of this book, the long-term effects of displacement and loss of life has resulted in significant negative effects for the lives of many survivors who live in resettlement communities, as well as their families. In attempt to recuperate local culture, and sovereignty, a handful of individuals left Pacux beginning in 1991 in order to pursue a new life on the steep slopes above the reservoir in Río Negro. Although respondents express the many difficulties they face in reestablishing the location; due in large part to the fact that the best, most arable land lies far beneath the water of the dammed Río Chixoy; they have been successful in re-inhabiting and recuperating the land, as well as initiating projects for their sustainability, such as the Centro Historico y Educativo, which houses a small museum and hostel. As confirmed during this research, the March 13 commemoration to Pocoxom, with its hundreds of participants—survivors and their families, as well as invited individuals—serves as a vital aspect of healing and reconciliation for the community.

As for the Chixoy Dam project post-1983, it has proved to be a financial disaster, even according to high-level officials in the INDE and World Bank (Witness for Peace 1996, 27). As early as 1987, reports indicated that "corruption of past military governments, engineering failures, and inflation left the dam two years behind schedule, and skyrocketed the original price from $340 million to $1 billion" (Bryson 1987). Sources report that up to $500 million dollars were lost due to

[9]It should be noted that since this was written some reparations have been paid to the affected families. More information on this complicated situation can be found in the epilogue.

corruption (Colajacomo 1999, 14). Besides the financial difficulties, which can be associated with nearly every mega-dam project worldwide (Scudder 2005), Chixoy is yet to deliver the electricity it promised, and as of 1999, had worked at best 70% of its expected power generation (Colajacomo 1999, 14). According to a 2004 report by the Network in Solidarity with the People of Guatemala (NISGUA), unexpected sedimentation rates will likely result in a much shorter anticipated lifespan, with a strong possibility of decommission within 20 years.

References

Arnauld, M. C. (1996). Relaciones interregionales en el área Maya durante el Postclásico en base a datos arquitectónicos. In: J. P. Laporte, & H. Escobedo (Eds.), *X Simposio de Investigaciones Arqueológicas en Guatemala*. Museo Nacional de Guatemala: Arqueología y Etnología.

Booth, A. J., & Walker, W. T. (1993). *Understanding Central America*. Oxford, UK: Westview Press.

Bryson, C. L. (1987). *Guatemala: A development dream turns into repayment nightmare. Touted high-tech dam saddles nation with high economic, social costs*. Christian Science Monitor.

COHRE (Centre on Housing Rights and Evictions). (2004). *Continuing the struggle for justice and accountability in Guatemala: Making reparations a reality in the Chixoy dam case*. COHRE Mission Report 2004. www.shr.aaas.org/guatemala/chixoy/chixoy.htm

CIIDH (International Center for Human Rights Research). (1999). *Draining the Sea: an analysis of terror in three rural communities in Guatemala (1980–1984)*. American Association of the Advancement of Sciences. Accessed June 1, 2008 http://shr.aaas.org/guatemala/ciidh/dts/toc.html

Colajacomo, J. (1999). *The Chixoy Dam: the Maya Achi' genocide*. The story of forced resettlement. Contributing Paper, World Commission on Dams. Cape Town, South Africa. Accessed March 10, 2008 http://www.dams.org/kbase/thematic/tr12.htm

Commission For Historical Clarification (CEH). (1999). *Guatemala: Memory of silence*. Report of the Commission for Historical Clarification Conclusions and Recommendations Guatemala, http://hrdata.aaas.org/ceh/report/english/

Davis, S. H. (1988). Introduction: Sowing the seeds of violence. In R. M. Carmack (Ed.), *Harvest of violence: The Maya Indians and the Guatemalan crisis*. Norman, Ok: University of Oklahoma Press.

Flynn, P., & McConohay, M. J (Directors). (2002). *Discovering Dominga*. Jaguar House Film.

Handy, J. (1984). *Gift of the devil*. Toronto, ON: Between The Lines.

Ichon, A. (1978). *La Cuenca Media Del Río Chixoy* (Guatemala). Cuadernos de estudio Guatemaltecos 3. Guatemala City, Guatemala.

Johnston, B. R. (2005a). *Volume one: Chixoy Dam legacy issues study, executive summary: Consequential damages and reparation: Recommendations for remedy*. Santa Cruz, California: Center for Political Ecology. http://www.centerforpoliticalecology.org/chixoy.html

Johnston, B. R. (2005b). *Volume two: Chixoy Dam legacy issues study*.

Jonas, S. (1991). *The battle for Guatemala: Rebels, death squads and U.S. power*. New York, NY: Westview Press.

Lovell, W. G. (1988). Surviving conquest: The Maya of Guatemala in historical perspective. *Latin American Research Review, 23*(2), 25–57.

Lynch, B. D. (2007). *The Chixoy Dam and the Achi Maya. Violence, ignorance, and the politics of blame*. Einaudi Center Working Paper. Cornell University. http://www.economyandsociety.org/events/conf_ethicsglobalization.shtml

Mann, C. (2006). *1491: New revelations of the Americas before Columbus*. New York, NY: Random House Inc.

Manz, B. (1981). *Refugees—Guatemalan troops clear Peten for oil exploration*. Cambridge, Massachusetts: Cultural Survival Quarterly, 5.3.

NISGUA. (2004). *Chixoy Dam/Río Negro Massacre Case*. Network in Solidarity with the People of Guatemala (NISGUA). http://www.nisgua.org/news_analysis/index.asp?id=2616

O'Brien, T. (2007). *Making the Americas: The United States and Latin America from the age of revolutions to the era of globalization*. Albuquerque, NM: University of New Mexico Press.

Osorio, J. T. (2003). *The Río Negro massacres*. Toronto, ON: Jesús Tecú Osorio & Rights Action.

Stewart, J. (2006). When local troubles become transnational: The transformation of a Guatemalan Indigenous rights movement. In: H. Johnston, & P. Alemida (Eds.), *Latin American social movements: Globalization, democratization, and transnational networks* (pp. 197–214). Maryland, USA: Rowman and Littlefield.

Witness For Peace. (1996). *A people dammed: The impact of the World Bank Chixoy hydroelectric project in Guatemala*. Washington DC, USA: A Witness for Peace Publication.

World Bank. (1978). *Staff appraisal report, Guatemala, Chixoy Power Project*. Regional Project Department, Latin America and the Caribbean Regional Office. http://web.worldbank.org/external/projects/main?pagePK=64283627&piPK=73230&theSitePK=40941&menuPK=228424&Projectid=P007189

Chapter 6
Lived Experiences: A Critical Analysis

Abstract The purpose of this chapter is to give a voice to those directly affected by the Chixoy Dam and Río Negro Massacres, as well as those who work closely with the victims. Through my analysis of interviews, observations, and participation in meetings, I identified three key themes that illuminate the perceived impacts of the Chixoy Hydroelectric Dam on the former residents of Río Negro. These include: (1) impacts from violence; (2) economic impacts; and (3) contemporary development. In order to better describe these broad topics, I recognized subthemes, which allow for a more detailed account.

Keywords Violence · Development · Progress · Trauma · Recovery · Cultural restoration · Land · Human rights · Genocide · Ethnographic research

6.1 Impacts from Violence

1. Loss of life

As explained earlier in this study, at least 444 citizens from the community of Río Negro lost their lives as a direct result of the violence from 1980 to 83 (Johnston 2005, 6). This number does not reflect the potential total number of victims who later lived in the resettlement community of Pacux, and continued to face military repression in the form of kidnappings, torture, and assassinations.

The issue of past violence and the loss of family members unified my respondents more than any other topic discussed. Although I do not claim to be an expert in the field of mental health, I found that most interviewees seem to have suffered from psychological trauma due to what they witnessed, and expressed a deep sadness, and loss, in growing up without their parents, brothers, and sisters. All of my respondents, both women and men, became emotional in some form or another when discussing what took place in Río Negro during the years of violence.

N. Einbinder, *Dams, Displacement and Development*, SpringerBriefs in Latin American Studies, DOI 10.1007/978-3-319-51511-3_6

On the night of March 13, 2009, at the commemoration of the massacre at Pocoxom, an informal interview with CRISTÓBAL, a survivor of the violence and resident of Pacux, turned into a very poignant telling of the story:

> My mother tried to escape, and I found her body, down there (pointing to the ravine that we were sitting near) sprawled out. She was shot and tried to escape. My wife had been carrying one of my daughters on her back. I found machete marks both on my daughter and my wife. I just found her there. She was cut in half…This is a very painful story but I thank you very much to come here to listen, because this is a very true story.

As should be expected, past violence has long-lasting effects on individuals, and on the entire community of Pacux, as well as the newly recolonized community of Río Negro. Young survivors, such as ROLANDO, who was born while his mother was in hiding, told me how "alcoholism and severe mental problems were common ailments due to what individuals [such as his own mother and stepfather] had witnessed and experienced." Other individuals, such as JULIÁN, believed that much of the youth gang violence in Pacux is "an effect of past violence. There is no one to teach them [orphans] to work…We were taught to work [in Río Negro], we grew up with our mothers and fathers."

Nearly, all the women I met in Pacux were particularly distressed when recounting what happened, and teared up, or openly wept, several times throughout the interview. Many spoke about the hardships of being orphaned, and believed that much of their economic and psychological difficulties were due to the fact that they had grown up without parents, and were raised either by another family member, or in an orphanage.

Individuals, such as MARÍA, a community activist in Pacux, stated that survivors were still "weak" as a result of growing up orphaned. And when speaking about potential financial reparations offered by the Guatemalan government, MARÍA, as well as other respondents, commented that there was no amount of money that would bring back a loved one, or fix the fact that they had endured childhood without their parents. ROSA places much of her past and present sadness on the loss of her mother, who died at Pocoxom on March 13th 1982:

> One suffers a lot when they do not have a mother. Because you don't have affection from no one…There is no happiness. There is no hope, because I can't see my mother, or visit her.

ROSA's story is not uncommon in that, unlike her mother, her father survived the massacre at Pocoxom as he was already in hiding, and believed that the military and PAC members would not harm the women, children, and elderly. Her father has since re-married, and problems arose between his new wife and ROSA. To exacerbate the problem, ROSA's father began to drink heavily after the violence while living in Pacux, before his eventual return to Río Negro. She claims he beat her repeatedly and hit her with his machete. She also stated that her stepmother would not feed or take care of her, as all her attention was placed on her own children. Again, ROSA recounts the sad memories of her childhood:

> I was nine years old when I met up with my father, and I went to live with him again. After that, I started to make weavings. And I didn't have time to make tortillas, and my stepmother didn't give me tortillas. I was hungry when I went to school. One teacher asked me why I was sick, and I told her, I am not sick, I'm hungry. Because my father does not feed me. Or my stepmother. I cried. Because one without a mother has nothing in this world.

While carrying out interviews in Pacux, I found that conflict between stepparents and their stepchildren was an all too common subject among those who lost just one parent to the violence.

In contrast to so many stories of sadness and despair, some interviewees (mainly men) spoke with relative ease about what took place, and talked of personal 're-habilitation' since the violence. References to Jesus Christ were frequent when interviewees discussed their own path toward well-being and reconciliation with what took place. I found that those able to speak with more controlled emotion about the violence generally lived in present day Río Negro, where the socioeconomic burdens are far less extreme than in Pacux. SEBASTIAN, a current resident of Río Negro and community leader, spoke of his recuperation:

> I am a bit recovered, because I have my house, and my wife has chickens, and dogs, and some ducks. My kids can study. My daughter GLORIA ANGELICA has graduated, and my other kid RODRIGO…I am really content, because he is studying in Livingston. I am struggling for them, and I have worked really hard for this. So now…I feel a lot better than before, because now we live directly in Río Negro, and also we have this center, and we can get a bit of money from it, and I can buy fish to eat…At least this is an advance in how I was feeling.

As will be discussed in greater detail in the section on 'development,' rehabilitation appears to be an easier process for those who are either working for reparations and justice, or have been able to return to a semi-subsistence lifestyle, such as those in Río Negro. It cannot be overstated how important it appeared for most of the men I spoke with to have their own land in which to cultivate a milpa (a traditional plot of corn, beans, and other assorted vegetables based on the local geography and climate), and return to a more traditional, Maya-Achí vision of life. And this, I believe, has aided some individuals considerably in the process of recuperation.

2. Physical displacement

Following the massacre at Pocoxom, survivors from Río Negro fled to the rugged mountainous terrain surrounding the Chixoy River basin, which began to flood shortly thereafter. Survivors left everything they owned in their homes, which were looted and eventually burned by soldiers and PAC members. Some fortunate individuals, such as SEBASTIAN, managed to arrive in time to "get corn and clothes and hide them in the mountains."

Most survivors lived from one to 3 years in the wooded gulches surrounding the basin, as a means of hiding from military and paramilitary forces. A small number of individuals made their way out of the mountains to Guatemala City, into the neighboring department of Alta Verapaz, and in very rare instances, the United States.

CRISTÓBAL, already 24 years old at the time of the violence, discusses life in hiding:

> We lived in small groups, all over, so we wouldn't be seen by the military. When we heard soldiers coming in from the Baja Verapaz side, we'd move over there [pointing up the mountainside], or from the Quiche side, we'd come over here...We slept under trees, with no water, no food. We would eat the middle of the palm stalks. We ate jocote roots, because there was no fruit. Uncooked. We fished at night. We ate it raw, without salt. We caught the fish with palm leaves and string. There is also a bean from a tree that we ate.

Many children perished due to the harsh conditions of living in the mountains, with limited food, water, and shelter. On May 14th, 1982, just 2 months after the massacre at Pocoxom, military forces tortured and killed an additional 84 survivors at Los Encuentros, located roughly five kilometers downstream from Río Negro. Days later in Agua Fría, another village within the vicinity, soldiers killed 35 children who were seeking refuge (Colajacomo 1999, 5).

Apart from those who fled to the mountains, PAC members took 18 children from the massacre site at Pocoxom to live at their homes in Xococ (Colajacomo 1999, 5). MARÍA, who was 14 at the time of her abduction, told me about the incident from her home in Pacux:

> I don't remember the day. I left in 1982. When there was a massacre in Río Negro. Two patrollers brought me...I lived in Xococ. I pretty much had no family. I remain without my family. The patrollers made me cry, made me sad. I remained without my mother, father, grandparents, brothers and sisters, no one. My father was killed in '80. My mother '82, on the 13th of February. My grandparents on the 13th of March. And after that month, I went with the patrollers. All the kids and old people and women were killed. I lived in Xococ. I lived there for about one year. I don't really remember that year. And for 25 years I have lived here in Pacux.

JESÚS TECÚ OSORIO, also taken in as a slave to Xococ, recounts in his book, The Río Negro Massacres (2003, 29), of life with his captor, Pedro Gonzalez Gomez:

> My life of uncertainty began the 14th of March. I didn't know what was waiting for me. I didn't know what kind of life I would have with Pedro Gonzalez. Soon enough I realized that I was a slave for this patroller. For over two years my life was suffering, pain and tears. I had no hope of being rescued by a relative or friend. I was a servant for that family...

Family members aided by the church eventually rescued JESÚS, along with other children captured from Río Negro. However, some victims, such as MARÍA's brother, died while in captivity.

In late 1983, after General Efraín Ríos Montt called an 'amnesty' for all refugees in hiding, most of the remaining survivors living in the mountains arrived to Pacux (though some originally ended up in camps in Alta Verapaz), to live in the houses built for them by INDE and the World Bank. Upon arriving, all persons had to present themselves at the military base, which was located across the street from the entrance to the village. SEBASTIAN recalls his eleven days locked up at the base:

> A man from my wife's community labeled us spies, so we were separated from everyone. We were taken to a small room, a latrine, with feces everywhere. We were tied up and chained to some pipes and separated into stalls. This is my story, but I could hear men in

> other stalls too. I was kicked and they grabbed my head and held a knife to my neck. "You're guerrillas," they said. "How many soldiers did you kill? Where are your weapons?"

After enduring many days of torture, SEBASTIAN told them that he had helped the guerrillas, thinking that by saying this, they would let him go. He spent another six days in the latrine, tied by his hands to the toilet. SEBASTIAN's brother, who had joined the guerrilla following the massacres, tried to escape from the base and was shot and killed. Upon release, a military commissioner told SEBASTIAN, "I spared your life, but others want to kill you." SEBASTIAN was soon forced into the local Civilian Patrol (PAC). CRISTÓBAL, who was also locked in the base for a total of twelve days, was finally rescued due to the help of a local Catholic priest, Father Melchor, who wrote letters to high-ranking commissioners and later had him released.

After moving into their homes, military soldiers and PAC members continually harassed survivors. As a 'model village' Pacux was under strict military control, and residents could not move about freely. For SEBASTIAN and many others, there was no other option but to leave Pacux, for the safety of himself and family, and to find work. He traveled to the sugar cane plantations along the south coast, and to the Petén, before finally heading back to Río Negro in 1991.

For the orphaned children, moving from one place to the next was common, at least until they reached an age where they could fend for themselves, or get married. GUADALUPE, an orphan who survived the March 13th massacre by hiding in the Río Negro church, describes what life was like in San Cristóbal, Alta Verapaz, where she ended up after living in the mountains with her brothers and sisters:

> We didn't have freedom to leave or pass. They [the soldiers] took care of us, with guns… Each morning, everyone, kids, old people, were gathered and talked to about the guerrilla, but we were kids, we didn't know anything, we were just poor kids. There, we suffered a lot. For one year we were there. And every morning, at five in the morning, everyone was awakened and checked on by the soldiers. They yelled at us, all of us, kids, men, women. They taught us to sing the anthem for the army. It was the law.

One respondent, EFRAÍN, who currently lives in Pacux, has a fairly typical story of periodic movement from one location to the next following his arrival to Rabinal from the mountains:

> When I was 15, in 1986, I was forced into the military, to work in Zone 1. I did it for three years, forced. After this I worked in a factory in Zone 13, Mixco [a suburb of Guatemala City]. I worked there for six or seven months. After this, it was very hard to survive, I had to buy everything, so I came here [Pacux]. I came to live in the house with my 13 brothers, then with my uncle. I searched for a way to survive. Some of my brothers went to the capital, to work on houses. I married in 1992, and we went to Río Negro…

A number of Pacux residents also migrated to the United States in search of work. Halfway through an interview with GUADALUPE I learned that her husband, also a Río Negro survivor, had been living in the U.S. for 3 years, with little contact. She was unsure where he presently resided, as he was just laid off from his job in Virginia. She appeared embarrassed, and a bit ashamed, by not knowing where he was living.

A fairly common subtheme that emerged from discussions regarding displacement from Río Negro is that of, what I call, 'the life before.' Although I tend to interpret these sentiments as more of an economic issue, the psychological impacts of such violent displacements—as well as the cultural impacts—are worthy to note in this section, and will be expanded on in my final discussion.

Below is MARÍA's response to my question about how life in Pacux compared to the life she remembers in Río Negro:

> Nothing is like living in your own community, when you don't need anything, when you have everything you need, without thinking. And now, we're thinking about so many things, it's not the same. Not like before, when everything was calm; there was no sadness, nothing. Not like now…there are a mountain of things that we're thinking, because, it is very hard, what happened. What happened here…

Throughout my time in the new Río Negro, references were made to important sites in the basin, many of which presently lie underwater. On two separate visits I went to Cahuinal, a Classic Maya ruin that surfaces only during the dry season. Those living in present day Río Negro spoke proudly about this location, as well as other ruins scattered throughout the basin. SEBASTIAN, as well as ÁNGEL, another long-term resident, often pointed out places of historical and cultural significance, such as meeting spots, other ruins, old corn fields, and places, where the women used to gather lime to make tortillas. There was frequent talk about particular locations, such as at Los Encuentros (also underwater), where people from all over the basin would meet to trade goods and dance.

Although other communities in the basin besides Río Negro, such as Chicruz, are repopulating in recent years, the rich cultural activities and trade that defined the region is almost completely lost due to the displacements and loss of suitable habitat as a result of the flood.

6.2 Economic Impacts

1. Loss of Land, Possessions, and Livelihood

> The truth is that, before, when I lived with my parents—before they built the dam—everything was good. I worked with my father; he had a business making artwork. We had materials, everything. We had peanut trees. There were many things we made: hammocks, petates, etcetera. When we lived in Río Negro, life was good. We had a small river, and fished at night. We had fruit trees: oranges, zapotes, and many other fruits.

The statement above, spoken by EFRAÍN from the small bible school that he operates out of his home in Pacux, is a common sentiment regarding what was lost due to the flooding and forced expulsion from their homes at Río Negro.

Respondents from both Pacux and the new Río Negro spoke about the bountiful harvests they had along the river's edge, and the free movement they had to trade goods at local markets. SILVESTRE, who was already middle-aged at the time of the violence, expands a bit on what he remembers:

> Before the violence, before the dam, we were better…We worked in agriculture. Almost all of us were campesinos. Each family worked their own land; planting maize, beans, squash, tomatoes, jocotes, and mangoes. Included in this was the fish. Because right now, where the reservoir is, we had a river, a small river. And there were many fish in it. We had our houses separated, because the place was big. So, we were good. We didn't buy wood, jocotes, and we didn't buy much maize. We sold beans and tomatoes…

Respondents spoke nostalgically about how the best land was now underwater—land that used to support the entire community. SEBASTIAN and ÁNGEL both told me how they used to have several species of fish, now locally extinct, and much better pasture for their animals.

Now that the valley is flooded, people speak of the limitations of travel. MARÍO stated how "before the dam, people could cross the valley freely, with cows and products to sell in Rabinal, but not now. We cannot pass because of the reservoir… and it is very difficult to get in and out."

When interviewees speak of loss, they frequently compare it to what they now have in both Pacux and the new Río Negro. Most of the respondents are resentful toward INDE due to the unfulfilled promises of free electricity, land for agriculture and animals, and job opportunities.

During interviews I was constantly reminded how, in Pacux, there was tremendous difficulty in finding wood, and land to plant crops—a sharp contrast as to how life was before in Río Negro, where both wood and land were plentiful and managed accordingly. MÁRIO told me how "the women were being accused as robbers because they were getting wood from the hills, as they always did. There are many who find work [in Pacux], and still must collect on private land." In asking MARÍA whom she blamed for her present economic struggles, she linked it clearly to what was lost, and not recovered (as originally promised by INDE and the World Bank):

> I place the blame on INDE because, before, we had no needs, not like now. Now, we have to buy wood, food, everything. But before in Río Negro, we didn't have to buy anything. Before, my father sold fish. But now, there's nothing. One pound of fish costs 15 Questzales (about $2 US).

MÁRIO also responded to the same question with anger and resentment toward INDE:

> There are problems with INDE because they didn't comply with their promises. When we were living here, they took advantage of the war of '82. They made us houses that we didn't want. And they only bought seven caballeros of land [in Río Negro there was roughly 22 caballeros]. And besides this, how many fruit trees lie under water? How many harvests did we lose in 30 years? How much do they owe us?

Respondents also blamed the patrollers who, besides massacring family members, took possessions from their homes. MÁRIO told me how "everything was stolen—the horses and cows—everything." SILVESTRE emphasized this point at several times throughout the interview:

> All of our clothes were robbed by the patrollers. They left us very poor, because there was money that we left in the houses. We left radios, important things. There, we suffered a lot.

According to the survivors, as well as human rights and development workers, it is clear that the loss of property, livelihood, and land, is a very negative economic impact on those displaced from Río Negro.

2. Increased Poverty

According to JOSÉ, director of ASCARA (Associacíon Campesina Río Negro 13 of March Maya-Achi), the community of Pacux is defined by its 'extreme poverty.' Even by Guatemalan standards, Pacux, and many of the other relocation villages set up by INDE and the World Bank, are desperately poor, without access to basic services, or opportunities to better their situation.

Lack of resources, economic opportunities, discrimination, and lack of outside help—which was guaranteed by the World Bank and INDE upon their forced displacement—are the dominant perceived factors regarding an increase in poverty and decrease in overall well-being since the construction of the dam. Again, EFRAÍN expresses his frustration:

> We have no electricity, no water—our basic rights. And before, the INDE offered us many things…they made us these houses, but these houses don't work. They need mending, and they were made 30 years ago. We are looking for a way to construct out of adobe, but there is no way, we don't have the resources.

CARLOS CHEN, co-director of COCAHICH, a development organization which works on behalf of all Chixoy Dam-affected communities, acknowledged that the level of poverty in Pacux is exacerbated by the extremely crowded conditions, where up to five families share one small home. Lack of services, such as water, intermittent electricity, and proper sewage systems result in slum-like conditions. Residents, such as GUADALUPE, told me how the water only comes on for 20 min a day, with no service on Sunday. Several women explained that there was not enough water to clean what little clothes they had.

Interviewees, such as ROLANDO, expressed that great difficulties exist for children growing up in such extreme poverty to break out of the cycle, due to the lack of education. JOSÉ from ASCARA gave his synopsis of the situation:

> The majority of families have no money, and if they don't have money, they can't send the kids to school…The kids can only go to primary school because they can't afford anything more. When the young people reach a certain age, they end up going to the capital to work, and there are also the youth gangs…

The increase in gang violence in Pacux is not unusual given the social and economic realities of the country (see Benson et al. 2008). A primary reason that the residents of Río Negro desire a secondary school in their community is so that their children do not have to live in Pacux, or Rabinal, in order to attend further schooling. ÁNGEL, from Río Negro, told me of a recent and tragic incident in Pacux, where his son was robbed and killed by a youth gang on his way to school. Nearly, everyone I spoke with blamed the local violence, at least in part, to the dire economic situations that persist. Again, I argue that this can also be linked to the psychological impacts of violence experienced among survivors.

Lack of employment is also a major topic among residents in Pacux. Due to the abrupt change from subsistence living, which was practiced in Río Negro, into a more 'modern' capitalist system, residents had to immediately generate income to provide for their needs. Upon arrival to Pacux in 1983, up until today, residents face discrimination by those who live in Rabinal, and are considered to be among the lowest rung of society. At several points, while talking with people in the market in Rabinal, to Ladinos and Indigenous alike, I was given stern warnings about going to Pacux alone, and even laughed at when I told them what I was doing. The residents in Pacux were very aware that this attitude exists, and state it as a reason none of them could obtain jobs in town.

As a result of so many factors, many survivors left to find work in the capital, in the plantations of the Boca Costa and Petén, and the United States. Survivors returned, and still do return, to Pacux, because it is there that they have homes (although many are structurally unsound) and a tight knit community of fellow survivors. However, finding a steady source of income is a challenge, if not impossibility, for nearly all of my interviewees.

Many women, and to a lesser extent men, have taken up weaving, which helps, but is inadequate. Because of the expense of materials, and the amount of time it takes to finish a hammock, placemat, or blanket, the net gains are low, and cannot sustain a family. JULIÁN, also a weaver, stated that he made only 25 Quetzales, roughly three dollars, for making a placemat, which required 2 days of work. HEIDI MCKINNON, a Peace Fellow for the Advocacy Project, a U.S. based NGO, is working on organizing a cooperative where local weavers can buy supplies at wholesale and sell direct to U.S. buyers. For women such as ROSA, learning to weave helps tremendously, as she can now work from home and take care of her young children, instead of constantly searching for work in Rabinal, or in the capital. "Always I thank God for having work in my house," she told me. "Sometimes people ask for a canasta (hand made bag) and I make it for them…I am content, because I am making money. I have to know people [to sell them]."

Yet still, ROSA spoke frequently of the desperate poverty that she and her family experience, and stated that if she came up with the money, she would like to go to the United States to find work. When I told her what I knew about immigrant experiences in the US—the constant insecurity and lack of work due to the recession—she stated that "there is just no other route. Right now, I have no money to get my kids what they need to study. They need shoes, uniforms. And I just don't have it…"

According to JOSÈ from ASCARA, most male inhabitants living in Pacux work intermittently as day laborers in Rabinal, while also planting a milpa during the rainy season on rented land. SILVESTRE, also a weaver, gave his opinion on how the current system is working out:

> Look: here, it is very difficult. I am talking for all the people, not just for me. Here, you can get work on some days, but for every day, there isn't work. And for those of us affected, there is nowhere to plant [crops for subsistence and to sell]. Here, there is very little land. We plant a little, and harvest. We only have two months for corn. Why? Because there is no land. We don't have land to get wood. The land you can see on the other side is the land of the Pueblo. It's private. But for the necessity of many people, they go to collect wood. They go, to rob wood. They rob because they need to, because they don't have money to buy wood. And before in Río Negro, they made petates [traditional mats they sold at regional markets], because the palm was close by…So, our lives here, now, are very difficult. Some women go to the village, to ask for work, to wash clothes. That's what they do, the women, to survive, because there is no possibility to live better.

Almost all the men I spoke with still desired to live once again as self-sufficient campesinos, or traditional farmers, and the women also supported this way of life for their families. They spoke of products they made and sold in Río Negro, and of the culturally rich process of planting and harvesting their milpa. Again, the main problem is land, particularly the lack of access. Any person who wishes to have a milpa must rent the land from a private owner. This rent cost often offsets the benefits of the harvest. SILVESTRE, along with a group of 20 men bought a sizeable piece of land to harvest their crops near the reservoir (6 to 8 hours away, by foot) that they work shortly before and during the rainy season (May to November). This purchase has proven to be successful, but it demands an incredible amount of resources and time to get to their farm. EFRAÍN rents land closer to Pacux where he grows jamaica (a flower which is used to make a sweet, traditional beverage) to sell at the market, as well as grasses for supplemental cattle feed. But, again, this strategy provides work for only a few months a year.

In Río Negro, where land is still fairly plentiful, residents stated that it is much tougher now to grow food, because the best land is underwater. As in times before, the harvest is crucial for survival, though presently it is not always sufficient. A variety of development projects—which will be discussed in further detail in the next section of this analysis—help the new residents of Río Negro to survive, and move away from the debilitating poverty experienced in Pacux. Projects such as the Centro Histórico, which hosts guests who pay nightly rates, and weaving projects supported by DOMINGA, a survivor adopted by a family in the U.S. during the violence, bring in additional income that residents share, and can be used to buy supplemental food.

However, depending on the standards by which we gauge poverty, one might perceive that the residents of Río Negro are still impoverished. From my own observations, I gather that the residents have enough to eat (at least during the time of my visit, at the height of the dry season), and live in adequate houses, but they lack the resources to provide proper sanitary conditions. Dogs, pigs, chickens, and other animals are inside homes, and the water from the reservoir, in which they catch fish and bathe, is obviously polluted from upstream sources (see Appendices I, F). Additionally, since the local school only serves children to the age of 10, those who wish to study further must relocate to Pacux or Rabinal, and typically live with family members there. Those who do not have the money, or desire, to continue going to school, remain in Río Negro with a very limited formal education—at least among 'western' standards.

Residents of Río Negro, as well as HEIDI, mentioned that since food supplies run thin for at least 2 months leading to the harvest season, residents must ration corn and beans, and have very limited vegetables, fruits, or meat. At the time of my visit, eggs and fish seemed plentiful, and I did not notice any signs of malnourishment in children or adults.

6.3 Contemporary Development

1. Social justice

> We have been struggling ever since [being displaced from Río Negro], against the dam—myself and Carlos—and we've all decided, that this is what we need to do with our lives. And if they kill us, well that's what will happen, because we are not going to stop…And what we ask from the government is not help, but reparations, and justice.

Although not everyone is as active in the movement for justice as CRISTÓBAL, who gave the statement above at Pocoxom on March 13th 2009, most interviewees from Pacux and Río Negro speak of the need for justice, and are involved in one way or another. A handful of men and women have, since 1984, risked their lives to fight for reparations and reconciliation for all the survivors. Organizations, such as ADIVIMA (The Association for the Integral Development of the Victims of the Violence of the Verapaces, Maya Achí) and COCAHICH, which are both heavily involved in the reparations negotiations and other major development objectives, were born out of this early movement.

According to Grahame Russell, Whose NGO Rights Action has been instrumental in the fight for justice in Rabinal, the initial exhumations in 1993 at Pocoxom "empowered survivors, by allowing them to take back the site." Since then, other sites have been exhumed, with the aid of the FAFG (Guatemalan Forensic Anthropology Foundation), in nearby communities, in Rabinal, and in the decommissioned military base adjacent to Pacux.

Those pressing for exhumations and justice in the form of reparations for damages caused, and criminal trials against the material and intellectual authors, are frequently threatened (to a lesser extent in recent years), by the military and paramilitary. The military labeled JESÚS a guerrilla throughout the early 1990s for his leadership in the exhumation of Pocoxom, and for his testimony against the PAC members who enslaved him and killed his brother, Jaime. In his book, The Río Negro Massacres (2003, 36), he explains the significance of exhumation day at Pocoxom:

> After the hard work of the anthropologists, we gave our families a Christian burial on the 24th of April, 1994. It was the saddest day for me because it was the last goodbye…I wanted to cry, I wanted to drown myself in alcohol but I didn't. I thought that these activities would help me forget the pain. This tragedy pushed me to continue the struggle.

> I spoke with people and shared everything that I had lived at the massacre with them. This was the only way I was able to alleviate pain, even though every word that I spoke reminded me of the anguish of the people who were massacred.

Throughout Guatemala, exhumations of massacred victims from the armed conflict bring forth considerable healing and empowerment among affected family and community members (Sanford 2008). Maya and Catholic (and Evangelical) services and ceremonies, as well as celebrations take place, to honor those who lost their lives due to the violence. In 1995, with the help of Rights Action, survivors built monuments in the Rabinal graveyard honoring those killed at Río Negro, and built a Maya-Achí cultural/holocaust museum.

Every year in Río Negro, on March the 13th, hundreds of survivors and their families gather at Pocoxom, to stay up all night and commemorate the 1982 massacre. As stated earlier, I had the privilege to take part in the 2009 memorial. Although people gave many emotional testimonies, I found the event to be more of a celebration, or even a social gathering, where survivors from both Río Negro and Pacux visit with each other, tell stories, and feast. Survivors such as CARLOS, and JULIÁN, gave speeches that helped me understand the importance of the event in terms of the collective health of survivors, and as a major step toward empowering the population.

In conjunction with the Centro Histórico (the collective cultural and ecotourism project in present day Río Negro), SEBASTIAN takes guests to Pocoxom, to tell his story of the massacre, and teach others about the experience. When I visited Pocoxom in 2008, with the UNBC field school, SEBASTIAN eagerly shared with us what happened, and ended our emotional visit with a statement, that I feel, shows the importance of the exhumation as it relates to the healing of his people:

> I thank each of you a thousand times for coming to this sacred place. Now we strive to rebuild our community and to give our children a future, to pass on our values to the next generation…I hope you all return one day.

Along with providing closure to survivors who lost loved ones during the violence, the exhumations also provide forensic evidence for lawsuits filed against ex-PAC members, as well as high-ranking officials.

In an unprecedented case, JESÚS, in 1998, testified against his captor, Pedro Gonzalez, who was later sentenced to life imprisonment. Another two PAC members from Xococ were also tried and sentenced. However, the intellectual authors of the massacre are still free, despite warrants out for their arrest (see Dill 2005).

Throughout the latter part of the 1990s, after the Peace Accords of 1996, which 'officially' ended the 36-year civil armed conflict, dialogue opened up between those affected by the Chixoy Dam, the INDE, the Guatemalan government, and the World Bank (Johnston 2005).

According to MÁRIO and CRISTÓBAL, negotiations were slow to transpire until September 7, 2004, when several hundred Chixoy-affected persons occupied the dam, effectively shutting it down. MÁRIO reflects on what this accomplished:

> Yea, we shut down the dam, but the next day we were labeled as terrorists by the newspapers. So now some of us are on watch lists. But now we have lawyers helping us, both internationals and nationals, to help us fight for what we are doing. Even though the government tried to terminate us all, we are stronger now, as we have all come together to fight. Many people have since left the country to speak to audiences and denounce those who did this to our community. The government has realized that we have all this international support, and when we have meetings we have observers and people from other communities who weren't affected…so the government has been forced to listen to us. The government has always accused us of things…during the violence they called us guerrillas, and now terrorists, but since the peace accords, things have been better. We have been fighting this for 30 years with the government…

Since 2004, the negotiation for reparations has been on the forefront of the social justice movement. Several outside collaborators, such as Barbara Rose Johnson of the California-based Center for Political Ecology, Rights Action, and the Organization of American States (OAS), have put together exhaustive documents that aim to document exactly what was lost due to the dam, and which promises were not carried out by the INDE. JULIÁN, who attended meetings between the INDE and the leaders of Río Negro in the early 1970s before the dam was built, and worked for the archaeologists who carried out preliminary research in the basin, gave me a firsthand account of what they were originally promised by the government, and the betrayal that later took place:

> They [INDE officials] entered the village and offered us many things, but nothing, or just a small bit of the promises were kept. The majority were not kept. One of the offers they made were to pay for the damages to the crops. INDE offered roads on both sides of the dam…They offered us trucks, but first they offered boats. To bring things from the market into town. They offered electricity. Because it was our land that gave them this, they offered it for free…But sadly, the dam was finished in 1982, and all that they promised before this, they didn't keep. And when there was a meeting at the INDE office, in San Cristóbal Verapaz, they kidnapped two of the committee members [from Río Negro]. And they took the act with the signatures and their promises. And the committee members turned up dead. The book of acts and promises disappeared…

In a fairly surprising move, the Guatemalan government, under the current Colom administration, agreed to provide reparations for Chixoy-affected communities. I find it important to note, however, that up to this point, both the World Bank and the IDB have failed to claim any responsibility for what happened, despite acknowledgement that the dam has since turned into a disaster, and their continued financial support throughout the violence—which they were apparently aware of (Witness For Peace 1996).

According to HEIDI, whose work for the Advocacy Project focused heavily on the negotiations, the reparation is a two-part process. Part one involves the most urgent needs for the relocation communities, usually involving projects initially promised by the INDE. Each community will receive its own unique package, based on the 'most dire' necessities identified by local leaders and submitted to the committee. New houses and school buildings, potable water projects, and electricity, are examples of guaranteed projects that some communities will receive in the next year.

The second part of the reparation process has to do with a development strategy for economic stimulation. According to JUAN DE DÍOS, director of ADIVIMA, the development strategy will focus on sustainability and self-sufficiency, as well as objectives expressed as important by villagers. JUAN expressed a need for the formation of cooperatives, both agricultural and in the craft making industries, and has worked with HEIDI for more than a year on generating a working plan.

Along with the recent developments regarding financial reparations promised by the Guatemalan government, an international trial has begun involving survivors of the massacre, with the Inter-American Human Rights Court system. During the time I was in Rabinal, CARLOS, MARÍA, and two other survivors flew to Washington D.C. to give testimony at a preliminary hearing. When I asked MARÍA what she wanted out of this trial, she stated the following:

> What we want in the court is justice…When you are a victim, there is no price. For this we are searching for justice…I am ready to rest, now, after 25 years of struggling. We have achieved the capture of one patroller, but this is an ex-PAC, not any soldier, or colonel—they are free. But hopefully we will resolve something later.

Like MARÍA, CARLOS also stated that the purpose of this trial is the capture of the colonel who ordered the massacre—the intellectual author of the massacres at Río Negro. However, CARLOS specified that he is also interested in obtaining financial compensation, similar to survivors of the massacre in the nearby community of Plan de Sánchez.[1]

The reparations process is complex, multifaceted, and at times, difficult to understand. Almost everyone I spoke with regarding the negotiations expressed some form of dissatisfaction or conflicting opinion about what is going wrong. Many people were not sure if they are on the list to be compensated because both of their parents were dead, and they lived in the house of their wife or husband. People expressed that ADIVIMA and COCAHICH are not consulting their opinion. After nearly 15 years of work by the Achí groups, the government has yet to comply with any promises for compensation, and interviewees seemed very skeptical that anything would actually transpire, despite the promising news from the capital and the ADIVIMA office. EFRAÍN, from Pacux, responds to a question regarding what he thought about the possibility of future reparations:

> Yes, I have heard of this. We have struggled greatly for this, but, there is a large confusion. You see, there is an office established, by the government, to repair the damages. There are reparations for social, material, and economic help, for the survivors of the massacres. In this colony, there was a study two months ago. The government came here, and said they would construct new homes. And also, there are some of us who went to the Inter-American court. I heard they began to construct homes in another finca [community farm], and that it will happen here. If we receive this, it will be a start to the damages of the dam and the massacres that happened. It has been very difficult to receive this. For the

---

[1]The massacre at Plan de Sánchez is well documented, and set a precedent in international cases relating to the massacres of the 1980s. Although those who carried out the massacre are still free, each affected person was given $25,000 by the Guatemalan State in 2004 (Inter-American Court of Human Rights 2004).

> Inter-American Court, most of us want to receive money, we want a little help, because we were orphaned. But for me, when they asked for my documents, and it showed that I was in the military, I was rejected. But they didn't know that I was forced. So I won't receive or benefit the same as my brothers. I won't receive nothing. Anyways...for now we haven't received nothing and continue to struggle. And there is the COCAHICH, and commission for the dam. And there are other organizations. But for three years we haven't seen anything. They talk, but there is nothing.

One area of sensitivity I found, which in turn has created internal arguments among survivors and their representatives, is due to the fact that the reparation agreements have become 'closed' between the government and the negotiators for Chixoy-affected peoples. As a result of this decision, it has been advised that discussions regarding compensation for the massacres and human rights abuses must be, for the time being, silenced. What this means, essentially, is that instead of the continued denunciation for what took place in Río Negro, and a demand for government compensation and legal action, negotiators are currently only asking for compensation of what they were promised by the INDE, without reference to the sticky subject of the massacres, and those responsible. This, they hope, will increase the chances of receiving funds from the government.

There is also presently a silencing of critique regarding the role of the World Bank/IDB, apparently due to the ironic fact that the World Bank will be providing the initial investment to the Guatemalan government for the reparations. According to HEIDI, a dialogue between the World Bank and Achí groups regarding possible future funding for stage two of the reparations campaign is also apparently underway.[2]

2. Perceptions of Development

> We have development projects, for some chickens. Ten chickens for each person. The best development idea would be for us to study with some compañeros (partners), to help us get a source of income. To lift us out. There is terrain, there is water, but money, there is none. We could plant corn, sweet corn, chilies. We could plant tomatoes, melon, watermelon, papaya. There's great land on the other side [of the river from Pacux]. We have dreams, wishes, but lamentably, we cannot achieve them. Some institutions, they could give us assistance, help our people. When there is more investment in the community, it is more tranquil. We search for a form in which to live, but we have no resources...

Despite the various responses I received from interviewees regarding their thoughts on 'positive development' in their communities, EFRAÍN, as quoted above, summarizes the dreams of so many in resettlement: to locate a source of income, obtain some form of outside investment, and return to the agriculturalist lifestyle so many of them practiced before the massacres and displacement.

The idea of an artisan cooperative was frequently talked about in both Río Negro and Pacux as a means to organize all the producers and find consistent markets for their goods. As stated earlier, HEIDI from the Advocacy Project works with

---

[2]As stated earlier, an updated version of what is taking place regarding the reparations agreements as well as the case at the Inter-American Court of Human Rights can be found in the epilogue of this book.

weavers from all the Chixoy-affected communities, and has applied for grants through the U.S. to get the project underway. By working at many levels, such as producing their own native cotton fibers for production, HEIDI sees this project as a way to involve people from all the communities, and help create a niche market for the goods produced.

Education is also an important topic among many interviewees, as so many of them were unable to study due to their displacement. According to many survivors, the education of their children was also a means to break the cycle poverty, and is seen as an opportunity and privilege. ROSA, who has three boys attending public school, partly with the aid of grants by the Unitarian Church, repeatedly stated the importance of her children getting a good education, and the exciting possibility that they might attend university. 'Thank God,' she told me, "that we are in Pacux, where there are teachers, and there are students who make it to university. Little by little things get better here…"

Both CARLOS, and JESÚS, who founded the innovative school Nueva Esperanza (New Hope) in 2003,[3] state that advances in education are among a list of 'positive developments' achieved since the early 1990's by survivors. And despite the fact that many children do not have the opportunity to study due to the economic hardships, people from Río Negro move back to Pacux in order to give their children the opportunity to attend Nueva Esparanza and the other nearby public schools.

Discussions with survivors about agriculture, either at the subsistence level, or as a means to provide income by selling at regional markets, are so abundant that they could fill an entire chapter. As stated earlier, the ability for survivors to have their own milpa in which to grow food for their family, and perhaps sell or trade other cash crops, was very important, and spoken about by nearly all of my respondents. Among some individuals, such as SILVESTRE, the only moment of real hope, pride, and excitement occurred when speaking of his plot of land (to which he holds title), and the harvest season. Many survivors indicate that a return to this way of life is a way to connect to the days of self-sufficiency and the cultural practices of their ancestors.

According to most residents, as well as what I observed, the major limiting factor for positive development in Pacux is the lack of adequate land. The INDE promised land equal to what they had in Río Negro, in quality and quantity, but what they received was an infertile, and far too small piece in Pacux. ROLANDO spoke of the inadequacy of the land they were given, and as a form of development, he wished for better land around Pacux (rather than land that was bought for them as

[3]Nueva Esperanza is a project designed to educate children whose parents were affected by the violence, particularly those who survived the massacres at Río Negro. There are currently three grades offered, from age 12–15, with an enrollment of approximately 120 students. The school, according to JESÚS, as well as current director Guillermo Chen, focuses on sustainability (primarily in agriculture), gender equality, and Maya-Achí history and culture. During my visit with the field school in May 2008, JESÚS informed us that, among many plans for the future, was a specialized nursing program.

compensation, located three hours away), and "support to the elders, who have knowledge about the Maya ways in which to grow crops." MÁRIO also commented on a return to "proper resource management and conservation" as a means of sustainable development, and is currently working with a local NGO, Madre Tierra, which focuses on organic methods and permaculture design.

In developing an alternative for residents in Pacux, at least for those who wish to return entirely to the Maya-campesino lifestyle, a finca called Saumax, located near Cobán in Alta Verapaz is currently being worked on by ASCARA, the reparations council, and Rights Action, as a means to accommodate residents. The land was originally bought by INDE in 1996, but due to the lack of funds to build houses, the piece remains vacant. According to JOSÉ, director of ASCARA, there will eventually be enough houses and space for 62 people, who are currently living in Pacux in extremely overcrowded conditions. There will also be opportunities to grow cash crops, such as cardamom and coffee will exist, due to the lush, tropical climate. CARLOS, despite his involvement in social issues in Pacux, is one of the residents interested in relocating to Saumax:

> Yes, I want to live there. I want to live there because there is a cattle project. Apart from this, there is a housing project from the state…It's a nice place people want to live because there are two harvests of maize. Right now there is corn; there are milpas…Right now in Pacux, there are six or seven families living in one house. So, if there are houses and water there, people will go. One or two of the families. To start another life in Saumax.

When I mentioned Saumax to some individuals, such as MARÍA and JULIÁN, they were not interested, because they could not imagine moving again, and having to become acclimatized to a new climate and environment.

As a means to work toward a development vision more aligned with the ways of their parents and ancestors, as well as escape the injustice which continued to occur in Pacux, three individuals—SEBASTIAN, MÁRIO, and JULIÁN—returned to Río Negro in 1991, and have since developed a community of twelve families.

The return, as they each described it, was extremely difficult, and many returnees went back to Pacux after a short period. For the first 2 years, they lived under nylon tarps with their families, and each month SEBASTIAN had to obtain signatures by military personnel in the department capital of Salamá, which proved that they had permission to live above the reservoir. They had to establish plots for growing their crops, and if anything was needed from the outside, they walked the 12 h to Rabinal. But as SEBASTIAN says over and over again, 'poco a poco'; little by little, life has improved.

Common responses to why people moved back to Río Negro included the free land that existed there to plant their milpas, as well as the high cost of living in Pacux. MÁRIO, who left Río Negro several years back in order to work for Madre Tierra, stated that families return because of the possibilities to have animals, wood, and access to fishing. "There are other possibilities to survive, in Río Negro," he told me. ÁNGEL mentioned that he enjoyed the fact that he was "free to move about in Río Negro, and it was safe." But, as stated by SEBASTIAN on multiple occasions, life in Río Negro is not easy, even now, and many people end up returning to Pacux because life as a campesino is constant work, and many from Pacux do not know (or enjoy) that way of life.

As well as the more practical, economic reasons for the return, SEBASTIAN, along with JULIÁN, mention the importance of moving back to their place of birth, and spoke of a deep connection with the land once inhabited by their parents and ancestors. As stated by JULIÁN, after I asked him why he returned:

> The reason we are here…INDE forcefully relocated us to Pacux. But I say, there are commitments…because if an animal from the mountain is relocated to a dry place, he dies, because he is used to a green land. And if an animal from the dry land is taken to the mountains, it is bad also, because he is accustomed to the dryness. And this is like us before, because INDE moved us like this. In Rabinal we couldn't live. Some are accustomed. And those who have become accustomed have jobs. But not all of us have jobs. Those of us who could not live there, returned to live here in our place. Because in our place, we were born. Because we knew that our parents and grandparents knew how to live here. Because in the pueblo (Pacux) we had to pay for electricity. We pay for potable water. There is land, there in Rabinal, but it is not communal land. And to get land from the mountains we had to steal. And to come upon the owner of this land, we were treated like thieves. So for these reasons we returned here. In the year 1991 on the 18th of May there were three of us from Rabinal who returned, because we were not accustomed to live there. And it is for this, that today we are here.

Despite the sentiments that JULIÁN and many other survivors feel for Río Negro, several of the residents I spoke with in Pacux did not want to return to Río Negro, as there are too many sad memories, and they have created a life in resettlement, and want to focus on the development of that community. Many also did not want to leave the free (albeit broken down) homes in Pacux, or the possibility to have their children attend higher education. It is also a significant financial undertaking to leave for Río Negro, where they must build another home, and quickly stake out land to cultivate. CARLOS feels that there is only enough good land in the new Río Negro to provide for the current resident families.

Development in Río Negro is an ongoing process and struggle, since the three men arrived nearly 20 years ago. SEBASTIAN, who is in charge of the Centro Histórico project, and is more or less the acting mayor of the community (what they call the COCODE), spoke in great detail about past and future projects that would bring income to the community in order for residents to purchase items that they could not produce on their own, such as medicines and clothes, as well as obtain basic services. Along with most of the inhabitants I spoke with in Río Negro, he was very satisfied with his work thus far:

> And for our kids, we now have a school. And for those of us who live here, we plant flowers, and have milpas: beans and maicillo [sorghum]. And we have no fear for what happened anymore, we are relaxed. We have potable water…and for this, I feel that we are advancing, and that with more support, from organizations and visitors, we will hope for more, not much, but just enough to survive, and to not suffer like we have suffered…

According to the description the survivor's gave me of the original village of Río Negro, as well as what was documented in pre-dam reports, it appears that the overall layout of Río Negro has once again developed toward a traditional

Maya-Achí way of life. Ichon's (1978) archaeological report, La Cuenca Media del Río Chixoy, explains how Río Negro had striking similarities to what he knew about pre-Hispanic Maya communities, with its layout of 'ranchos' or family compounds, surrounded by their milpas, as well as an assortment of other food producing trees and plants grown around the house. Ichon wrote of each family having one to three head of cattle, with fishing, chickens, and pigs as a common form of animal production. My observations in Río Negro, and the visions of development expressed to me, are identical to Ichon's description of the pre-dam community.

SEBASTIAN, as well as others with whom I spoke, are happy to live in Río Negro, because of the life they could offer their children, as campesinos loyal to their Maya-Achí heritage, and away from the negative elements of Pacux, and the capital. Residents are also content to be able to provide a life for their offspring, where if they wanted to study away from Río Negro, that they would have that opportunity, and hopefully return at a later point. JULIÁN stressed the importance of having a home, for his family and all the survivors of the massacres, in which to develop as they see it fit, and hopefully never again be uprooted.

The successful development of Río Negro could not have been achieved without the help of outside organizations and individuals, namely that of VICTOR LINDENMAYER. VICTOR is a German development worker who by chance heard the Río Negro story, back in the late 1990s, while working for Servicio Aleman, an NGO based in the capital. In the beginning, his organization funded small projects, such as the purchasing of boats, fishing nets, and provided small grants for residents to invest on things as they pleased, such as a church.

Since 2006, however, VICTOR has embarked on a rather large endeavor, the Centro Histórico, which he expressed to me, was 'his dream project.' Located on a scenic bluff over the reservoir, the Centro is part museum, part hostel, and meeting hall. On the slope above the main building, herb and vegetable gardens, a tree nursery, as well as several traditional houses exist to educate guests on how the Maya-Achí once lived, before the dam. VICTOR sees this as a project of "empowerment, and healing," where residents can earn money for goods they cannot grow or make, as well as "educate both national and international guests about what took place during the violence, and provide a unique cultural experience that is both remote and comfortable." He spoke of many future ideas for the center, such as hiking and other nature tours, and archaeological explorations, while always maintaining the focus of Maya-Achí history and culture.

SEBASTIAN, who is the current director of the Centro Histórico, explained how the project was started, and currently run:

> VICTOR saw that all the small projects were going well, and one of the dreams they had was to build a center for visitors, so VICTOR began to apply for money to help build the center here. It was hard to build this place, because we had to carry up everything from the water; the rocks, the wood, everything, cement blocks. It rook one year to build this, how it is now. The whole community is content to have this here, to receive groups, even though it's not that good, we are happy. Everyone takes turns working, the women and the men, all twelve families…and it really helps the community, because we divide all the money we

> make between us, to buy things that we need. It's really helpful for the women, because they all take turns doing things, like washing the sheets, or cooking, and they get paid for each job they do. And for the men, if they go get the lancha [small boat], or whatever they do, they get paid for that job. The people of Río Negro are in charge of the project, not the Germans, and when there is money left over, it goes back into the project.

When I asked others in Río Negro about the Centro Histórico, both men and women, all respondents seem content about how it was being run, and its purpose to bring guests and income. During my time in the community, I stayed at the Centro, and at times, small groups of students, mainly from the U.S., came to use the facility. Although many individuals from the community are involved with the various tasks of accommodating guests, it is always SEBASTIAN who speaks with the guests, and arranges any trips. The women show up to cook food, but rarely speak with the guests (it was under my assumption that many women in the community do not speak much Spanish). Due to the spread out layout of the community, and the constant work required for a campesino lifestyle, many people seem to have little involvement in the project.

Although the Centro is fairly new (officially starting to receive guests in March 2008), it still has a ways to go, according to VICTOR and SEBASTIAN, before it will bring in enough income to the village to offset the general upkeep, maintenance, and construction of other unfinished projects associated with the overall plan. VICTOR spoke of the need for more advertising, and of his hopes to bring in long-term guests and volunteers to learn how to weave or farm.

Future development objectives, expressed by my Río Negro interviewees, include projects that will bring in cattle and a better system to sell their weavings and hammocks. Perhaps at the forefront, however, is the construction of a secondary school. Everyone I spoke with expressed the difficulty in finding and keeping teachers. Currently, only one teacher travels back and forth from Rabinal, every four days. The lack of funds necessary to build another structure is an issue. Part of the reparation demands for the community is to further develop this idea.

As for the cattle project, respondents make continuous references to the cows they had before the dam, how they were stolen by the patrollers, and that at the present time barely anyone has cattle. They also make reference to the poor quality pasture, especially during the dry season. The grazing potential in the basin is low due to the aridity and steepness and areas where animals graze, I observed excessive erosion (see Appendices I, B).

I hoped to learn more about community members' perspectives on future potential mining in the region. Mining currently dominates the 'development' debate in Guatemala, and some preliminary explorations are evident in the region; a topic which I will return to in the final chapter. Most respondents did not have any idea about this situation, and did not comment. SEBASTIAN knew that geologists had come to scope out the possibilities, but seemed unconcerned. However, those involved in the social movement were well aware of these prospects, and stated their strong opposition, even if it did bring in much needed income. MÁRIO and CRISTÓBAL are both involved in local struggles against potential new mining operations in other districts of the department, as well as the present struggle against new hydroelectric dams in the country. MÁRIO commented on this matter, while speaking to a group of high school students at the Centro Histórico:

> There are other people from the community that are involved in the world struggle over dams. These exchanges are very important, because there are other projects in the country right now that may happen, and we don't want this to happen again. Right now there is a project in the Petén that has been stopped due to the resistance, and we have been successful in this due to the collaboration. There is importance to maintain this solidarity, so the government doesn't go on building more dams without fulfilling their promises to people who have already lost their lands...

6.4 Conclusion

The analysis presented in this chapter intends to give a direct voice to those affected by the Chixoy Dam. Though I developed the themes and subthemes evident in the interviews, I consciously let the voices of my interviewee's remain the focus of this chapter. In connecting this material to the literature I provided in the previous chapters, I aim to draw conclusions in the next chapter, which consists of a discussion of my overall findings.

References

Benson, P., Fischer, E. F., & Thomas, K. (2008). Resocializing suffering: Neoliberalism, accusation, and the sociopolitical context of Guatemala's new violence. *Latin American Perspectives, 35*(5), 35–58.

Colajacomo, J. (1999). The Chixoy Dam: The Maya Achi' Genocide. The Story of Forced Resettlement. Contributing Paper, World Commission on Dams. Cape Town, South Africa. http://www.dams.org/kbase/thematic/tr12.htm

Dill, K. (2005). International human-rights and local justice in Guatemala: The Río Negro and Agua Fria trials. *Cultural Dynamics, 17*(3), 323–350.

Ichon, A. (1978). *La Cuenca Media Del Río Chixoy* (Guatemala). Cuadernos de estudio Guatemaltecos 3. Guatemala City, Guatemala.

Inter-American Court of Human Rights (IACHR). (2004). Case of the Plan de Sánchez massacre vs. Guatemala: Judgment of November 14, 2004 reparations. http://www.corteidh.or.cr/index.cfm?&CFID=666614&CFTOKEN=69520161

Johnston, B. R. (2005). Volume One: Chixoy Dam Legacy Issues Study, Executive Summary: Consequential Damages and Reparation: Recommendations For Remedy. Center for Political Ecology, Santa Cruz, California. http://www.centerforpoliticalecology.org/chixoy.html

Osorio, J. T. (2003). *The Río Negro massacres*. Toronto, ON: Jesús Tecú Osorio & Rights Action.

Sanford, V. (2008). From genocide to femicide: Impunity and human rights in twenty-first century Guatemala. *Journal of Human Rights, 7,* 104–122.

Witness For Peace. (1996). *A people dammed: The impact of the World Bank Chixoy Hydroelectric Project in Guatemala*. Washington D.C., USA: A Witness for Peace Publication.

Chapter 7
Final Discussion and Conclusion

Abstract At this point I would like to return to my original objectives for this study and discuss the conclusions that might be drawn from the analysis and corresponding literature review. As stated in the introductory chapter, my aim for this research is to illuminate the 'lived experiences' of Río Negro citizens affected by the development of the Chixoy Dam, and place their stories and continued struggles into both the national and international development discourse. In the end, after reviewing key bodies of literature and synthesizing my discussions and observations from the field, the primary question from this research remains as it was in the very beginning: What can we learn about 'development' through this very specific, yet seemingly emblematic, case study?

Keywords Development · Progress · Neoliberalism · Violence · Conflict · Indigenous peoples · Mining

While there exists a wide range of avenues in which to grapple this question, from Chixoy's symbolism in hydroelectric dams worldwide, as well as the similarities it shares with other massacres that occurred in Guatemala throughout the 1970s and 80s, my own experience and exposure to issues in both Guatemala and North America result in an approach that integrates both local and international themes, focusing primarily on the divergent perceptions of land and 'development' existing between the dominant capitalist model and that of Indigenous and/or marginalized peoples.

My persistent attempt to connect the incidents at Chixoy to the recent surge in mining development throughout this book comes as a result of my time spent in San Marcos, near the vicinity of the Canadian-owned (Goldcorp) Marlin Mine, as well as in El Estor, Izabal, where tensions over the past 10 years escalate due to former Canadian-owned Hud Bay Minerals claim to Indigenous-occupied lands (see Imai et al. 2007). As mentioned in previous chapters, the forced evictions, contrasting perceptions between Ladino/international capitalists and local Indigenous residents, and the neoliberal concepts of 'development' expressed in these projects, share

N. Einbinder, *Dams, Displacement and Development*, SpringerBriefs in Latin American Studies, DOI 10.1007/978-3-319-51511-3_7

many relevant characteristics with the events from Chixoy (as well as a myriad of other projects over the country's history), and will be further expanded on in the following discussion.

In locating the events from Chixoy and Río Negro within the global context, I reiterate the significant influence of the dominant development model, which continues to displace and discredit other ways of thinking about 'progress,' and 'development' around the world (Escobar 2003; Davis 2009). Because of my exposure to the intensifying resource extraction activities in British Columbia (where I was based for my graduate studies) and the controversy that surrounds it, I make a comparison to this specific location towards the end of this discussion. However, I acknowledge the fact that two very different political contexts exist in Guatemala and Canada. My connection between these two regions are simply meant to reiterate that the same processes of exploitation, and contrasting views of land and how it should be developed, exist in various regions around the globe, and are informed by the same free-market, capitalist models.

To begin this discussion, I return to the Guatemalan context, where the events at Río Negro mirror that of countless other massacres and forced evictions resulting from development activities imposed by the elite and their military counterparts (see Carmack 1988; Jonas 1991; Handy 1984; Grandin 2004; CEH 1999). Violence, in the form of intimidation and physical acts, is a tool continuously employed by the oligarchy to ensure ample land and labor for their capitalist endeavors since the 'liberal' era of the late 1800s and into the 'neoliberal' present. During the civil armed conflict of 1960–96—and particularly that of 1978–83—violence became a common strategy employed by the military to remove the supposed communist insurgency (Manz 2005, 21). Using this excuse the population witnessed numerous assaults on Indigenous lands as a means to continue an aggressive national plan for modernization, as well as pacify the growing 'subversive' populations in the countryside (see May 1999). As might be imagined, the consequences from these events vary widely across landscapes and communities due to the diversity of terrain, level of economic activity, and resistant response (Lovell 1988). Through my ethnographic research in Río Negro and Pacux, I aimed to provide detail and voice to one of the many hundreds, if not thousands, of communities and individuals who fell victim to this prevailing system.

As stated in the previous chapter, those who survived such horrific events as the massacres at Río Negro continue to suffer, both emotionally and psychologically; and given the situation in Pacux, their economic hardships and cultural survival appears equally trying. Due to the loss of land and family, the social and cultural fabric on which the community depended was torn apart, and the repercussions will likely spill into future generations. As described, those who return to a more self-sufficient, traditional agrarian-based lifestyle, as they aspire for in the new Río Negro—or are working towards with alternatives such as Saumax in Alta Verapaz—appear more content, and less economically and 'culturally' impoverished than those attempting to survive within the bleak economic condition endemic to Pacux. By placing this assumption into a broader context—one that might aid us in understanding the larger implications of such long- lived policy and its effect on culture and community—I continue returning to the issue of land, and how its perception, utilization, and control has so effectively dictated the political, economic, and cultural outcomes within the country.

In many ways, the 'clash' that occurred in Río Negro, between the Indigenous campesinos and the INDE/army, symbolizes the divergent perceptions of land and development existing in Guatemala since the Spanish conquest. The INDE/army responsible for the murders and illegal displacements in Río Negro clearly represent ideologies and principles of the Ladino landholding elite, with ideas for 'progress' validated by the Western model (see Wright 2004), and in this case, the world's most influential development bank (Johnston 2005a, b). The Maya-Achí farmers of Río Negro represent an enduring class of Indigenous peoples, whose identity and livelihoods depend on the land in which they farm, fish, trade, and dwell with their families (Lynch 2007; Hamilton and Fischer 2005). Like many remote Indigenous communities, Río Negro's isolation allowed for a "culture of refuge," or "fugitivism," as described by Lovell (1988, 34) in regards to Maya communities under colonial rule, and as a result, self-sufficient and culturally intact livelihoods flourished and social networks were strong (Ichon 1978)—and worth defending, as expressed by the resistance.

Broadly speaking, Lovell (1988, 27) defines this universal attachment and "persistent defense," of land, and community, as "fundamental to the maintenance of Maya identity." Although the Maya-Achí of Río Negro lived along the Chixoy River for just a few generations [between 150–300 years depending on the respondent (see Osorio 2003, 1)], the strong collective resistance at the time of dam construction, and at other points[1] qualifies this statement, as do the actions presently carried out by survivors and their families. Individuals such as JULIÁN, one of the original returnees to Río Negro, stressed the symbolic importance of this persistence on several occasions, as well as the violent clash that followed:

> Our elders didn't want the construction, because they knew we would be left without our land, without anything. And so where the dam occupies, there lived our ancestors, there they planted… So, our elders went to protest. And what did the company say, the INDE, when they arrived to protest their rights? They said that they are guerrillas. They are guerrillas because the guerrillas also protest their rights. So, we need to get rid of them. And so we left all of our things…

Perhaps more telling than the statement above, however, is JULIÁN's opinion regarding the effects of displacement, and the idea of yet another move to Saumax:

> I make a comparison to the cutting down of a tree, because INDE took us out of our native land. Forcefully. We were forcefully taken out from Río Negro and planted in Rabinal. And in Rabinal, we couldn't take hold of our roots. Because it is not where we

[1]Previous examples of resistance to a Ladino-led 'invasion' of Río Negro land is stated in Osorio's documentation, The Río Negro Massacres (2003, 3). In his remembrance of stories from this time (around 1970), Osorio quotes the elders: "Let's fight to win the land so that the only owners will be our grandchildren" (2003, 3). According to this account, Río Negro residents, as well as members of the neighboring villages of Xococ and Pajales, successfully defended their land, with the help of lawyers.

> were born. And now, what they say, is that you will be taken out from Rabinal, and planted in Alta Verapaz [Saumax]. And what will happen, is that they won't be able to locate their roots.

Such profound connection—and commitment—to the land was demonstrated by the initial resistance, and confirmed once again by residents of the 'new' Río Negro, and struggling community members in Pacux.[2] While the issue economic practicality was certainly a factor in bringing survivors back to the land above their old village site—as resources such as wood, water, and fertile ground are far more abundant than one would encounter in Pacux—their continued dedication to the land, and practice of a more traditional Maya-Achí existence, is highly representative of the resiliency shown by the Maya people as a while. Through tremendous adversity, individuals such as SEBASTIAN—whose mere existence is amazing enough—risked their lives in order to return to the land of their birth and continue the rituals that define their culture: the tending of the milpa, the use of Achí language and other cultural practices, stewardship of the land, and development of community—as they envision it.

It is through comparable diligence—along with a near constant resistance (Montejo 2005, 27)—that Maya communities have endured nearly five hundred years of economic and racial oppression, assimilation attempts, state-sponsored terror, and the present phase of globalization (Lovell 1988; Manz 2008). Although respondents from Río Negro did not mention to me an affiliation with the greater Pan-Maya movement active in Guatemala (see Warren 1998; Montejo 2005; Fischer and Brown 2007)— nor did I mention it at the time—their tactics for cultural perseverance share characteristics with this larger theoretical and political entity.[3]

Despite the wide-ranging interests the Pan-Maya movement currently represents, its main objective of cultural survival and self-determination speaks directly to the activities of massacre survivors in Río Negro and Pacux. As stated by Pan-Maya activist and academic Montejo (2005, 156):

> With the recharging of our Maya cultural traditions and the emergence of a Maya unity with plural identity, we are ensuring that the voices of our ancestors and elders are becoming louder and louder again. We must struggle to maintain our life where our placentas have been planted, since the land is essential for our lives and provides our communal spirituality.

---

[2] It is important to note that although most survivors that I spoke with (living in Pacux) do not wish to return to Río Negro, for various reasons (as discussed in the analysis), nearly everyone had aspirations to return to a more self-sufficient traditional lifestyle, particularly in regards to growing their own milpa.

[3] As described by anthropologists Fischer (1996, 32–49) and Warren (1998, 108–115), the all encompassing, or modern, Pan-Maya movement began soon after the worst years of the civil conflict (1978–1984), as a reflection of the space offered to groups pursuing cultural revitalization, and resisting the oppression Indigenous communities faced throughout the country. In essence, the movement is one that attempts to connect and unite all Maya ethnic groups through language study, Maya academic scholarship, a reinvigoration of Maya cosmology, and solution-based resistance to the unequal power structures that dominate the development discourse, such as the poor distribution of land.

Acknowledging that development activities initiated by the state historically isolate and discredit Maya tradition and livelihoods, Pan-Maya groups offer alternative strategies that incorporate language and cultural revitalization as integral components, rather than the traditional capitalist, 'Western-based' approaches (see Fischer 1996, 225). In addition to the act of re-inhabiting land which was legally theirs, and moving toward a cultural rehabilitation and rejection of the dominant economic paradigm, the Centro Histórico tourism and cultural project offers a living example of the dynamic Pan-Maya ideology, as it incorporates history, culture, and language, while merging sustainability and innovation. It also represents the ever changing, and often survivalist tactics broadly representative in Maya culture. As stated by Gutierrez (1994, 1) on the dynamic nature of Maya cultural endurance:

> Fifty years ago, people were prophesizing the progressive ladinization of the indigenous people. Today, these predictions have collapsed. The Maya have a startling—seemingly limitless—capacity to absorb elements of foreign cultures and transform them according to their own code of understanding of the world and of the balance among its elements.

Gutierrez (1994, 1) goes on to recognize that aspects of Maya dress, diet, and religion are subject to the importation of ideas and strategies born outside Maya lands. Other writers on this subject, such as Fischer (1996, 32), note that "Maya activists necessarily seek their inspirations from the past—the continuity of their cultural tradition gives them the moral justification to call for Maya revitalization—and yet their philosophy is forward looking." Without abandoning their traditional livelihoods, such as farming milpa, and making tortillas and palm woven petates, the people of Río Negro pursue economic independence through the Centro Histórico project, along with the pursuit of international markets for non-traditionally made crafts such as blankets and hammocks. Ideally, these activities will aid in Río Negro's survival, while also giving residents the opportunity to work exclusively in their communities rather than migrate to large fincas as their ancestors did.

Other Pan-Maya initiatives such as education reform (see Montejo 2005) are also represented among the Maya-Achí, particularly in Rabinal, where it is easier to establish progressive schools for the children of massacre survivors. As noted in previous chapters, the establishment of Fundación Nueva Esperanza is an example of cultural perseverance by focusing on traditional Maya-Achí culture and history, while also offering practical skills for those who wish to escape the poverty inherent to wage-labor employment.

Despite these relatively successful attempts by survivors to maintain culture—through land re-habitation and other Achí-centered development activities—a fundamental difference of ideologies still exists between the powerful Ladino elite class, and the generally power-deprived Maya campesinos. And as long as issues concerning land/resource exploitation and unequal distribution remain unresolved at the national level, tensions are likely to further develop, thus affecting the overall success in Río Negro, Pacux, and the myriad of other Indigenous communities throughout the country with similar objectives.

In continuing with the discussion on the significance of land, and its relationship to development, I find it crucial to bring to light a more international perspective, as discourse and policy at this level inevitably influences the debate in Guatemala, both in a historical and contemporary context. As mentioned in the literature review, development, particularly in the post-WWII context, has been on a whole detrimental to Indigenous peoples worldwide (Power 2003; Bodley 1990; Davis 2009). As stated by Tauli-Corpuz (2008, 2), an indigenous activist with the Tebtebba Foundation in Bagulo City, Philippines:

> In the post-colonial era when nation-building was the political and economic agenda of newly independent states, development and modernity provided the justification for the re-colonization of our territories and attempts to obliterate our identities and cultures. The homogenizing and centralizing powers of modern states has led to the destruction of some indigenous societies and cultures, which is one of the great tragedies of humankind.

Modern industrial growth and material accumulation as global indicators of 'progress,' (see Wright 2009), along with the increasingly divergent views of land and its meanings (see Bodley 1990), are two important factors in the systematic upheaval of Indigenous peoples from their land—which, as mentioned, adds significantly to the process of cultural and environmental degradation (see Windsor and McVey 2005; Escobar 1995). Ideologies from this paradigm are harmful to Indigenous societies not only because they systematically discredit other ways of thinking about development, but also in the way they legitimize the excessively consumptive habits of 'developed' societies, which in turn leads to physical displacement as land becomes apprehended for export-driven production, as well as infrastructure projects to generate the energy necessary to continue such activities (Handy 1984; Galeano 1973). To reiterate the effects of this global system in a Guatemalan context, I turn once again to Montejo (2005, 26–27):

> The impact of Western economies, including feudalism and capitalism, in the entire Maya region has been another such external force [in addition to the importation of religion]. As traditional Maya economic systems have been replaced or modified, the impact of capitalism can be noticed everywhere. The new economic models have provoked ruptures in traditional lifestyles. Different modes of expropriation and exploitation of native land, from encomiendas and haciendas to plantations, submerged indigenous people into dependent economic positions…

While the activities of the multinational corporations cannot be disqualified from its heavy impact on foreign (underdeveloped) economies, the awesome power and influence of the International Financial Institutions on the importation of 'Western' capitalist ideologies and policies into nations such as Guatemala cannot be overstated (Power 2003, 75; Witness For Peace 1996). Although it would be difficult, if not impossible, to hold the World Bank and IDB accountable for the violent evictions and genocide that took place at Río Negro, it is instructive within their reports on the project, and that of the INDE and the foreign contractors, in how the Eurocentric—and openly racist—attitudes could have legitimated the discrediting of Maya-Achí livelihoods and their lands, thus justifying the use of violence. After gaining access to original feasibility studies generated by the development agencies, Cornell University researcher Lynch (2007, 21) offers a potent reflection of her findings:

> In this [project] culture, modernization was a goal to be achieved, non-commodity agriculture and cultural production was systematically undervalued. Language skills were not seen as particularly relevant to the data gathering process, which in any case favored the collection of quantitative over qualitative data. Economic calculations obscured the impacts of construction on peoples' whose economies are not fully monetized and integrated into the national economy; analyses systematically undervalue crops and livestock produced for subsistence or for local markets, nor did they take into account the importance of resource complementary in rural livelihood strategies.

Lynch's (2007) report is particularly important, not only because of her access to key documents, but because of the systematic manner in which she describes how a project like this came into fruition. "With one exception," Lynch (2007, 26) concludes "the project literature portrayed residents of the project area as a scattered, unidimensional group of backwards folk resistant to change."

However, demonstrative these statements are in the ways in which development projects are perceived and carried out, it should be acknowledged that these drafts were written over 30 years ago, in a different time and space. Policy within institutions such as the World Bank has apparently evolved to accommodate Indigenous views, independence, human rights, and involuntary resettlement (see World Bank 2005). In Guatemala the 'war' between the leftist guerrillas and the state is officially over, and with the Peace Accords of 1996 came the ratification of ILO 169, protecting the rights of Indigenous peoples (ADISMI 2007, 4), and opening of space for organized dissent (Fulmer et al. 2008). However, despite these good intentions and movements towards a more equal society (or at least the façade), I am left wondering: Have attitudes concerning 'development' really evolved as some proclaim? Are ideologies regarding land, and the contrasting views of those who remain powerless against the tide of modernization and 'progress,' accounted for in the grand scheme of globalization and Western-driven, neoliberal development? In my attempt to answer these questions, I look to the development ideology, and activity, not just in Guatemala but also in the Global North, which continues—perhaps even more tenaciously—to decimate culture and critical habitats in the name of economic growth and prosperity.

As mentioned in previous chapters, resource exploitation (mining, oil and gas drilling, energy producing facilities) is increasing in scope through the deregulating freedoms of neoliberalism, with conflicts paralleling that of Chixoy taking shape across many regions of the world. In Guatemala, where struggles over land are seen as a major instigator of violence since the removal of Arbenz in 1954 (Handy 1984; Lovell 1988), mining offers the most contemporary threat to Indigenous, agrarian-based communities. As stated by Holt-Gimenez (2007, 23):

> Over the next decade, agrarian struggles over land in Guatemala's Western Highlands may well be eclipsed by indigenous movements against mineral extraction. Both are essentially struggles for livelihoods and cultural survival. Indigenous resistance to territorial

> restructuring is not only crucial for confronting capital's threats to indigenous survival, it is necessary for embarking on purposive action for the "restructuring from below" that reaffirms indigenous space and place in the face territorial restructuring "from above."

In other regions of the world the possibility for huge monetary gains by transnational corporations has lead to similar exploitation, resulting in heightened tensions over land, rights, and divergent perceptions of livelihood. In Northern British Columbia, for example, where resources are still abundant and the population relatively sparse, the increase in industrial development is staggering, while those opposed to the imposition of such activity—primarily consisting of historically repressed First Nations bands—are offered limited power to defend their sacred, traditional lands from corporate and provincial interests (Alfred 2009; Union of British Columbia Indian Chiefs 2005; Harris 2003). With the price of minerals such as gold at an all time high, the race for licenses by the corporations to dredge, drill, remove, and pulverize the earth in the name of quick profit seems unstoppable (LaPlante 2008). And through a host of neoliberal reforms at national and provincial levels, corporations are now able to move through a once stringent regulation process with greater ease (Young 2008). On reflection of one recent proposal made by Royal Dutch Shell, to extract coal bed methane from the fragile headwaters of the Stikine, Skeena, and Nass rivers in northwest British Columbia, a land revered by local Indigenous groups, Davis (2009, 118–119) explores the absurdity of this form of 'development,' and what it represents:

> Environmental concerns aside, think for a moment what these proposals imply about our culture. We accept it as normal that people who have never been on the land, who have no history or connection to the country, may legally secure the right to come in and by the very nature of their enterprises leave in their wake a cultural and physical landscape utterly transformed and desecrated. What's more, in granting such mining concessions, often initially for trivial sums to speculators from distant cities, companies cobbled together with less history than my dog, we place no cultural or market value on the land itself. The cost of destroying a natural asset, or its inherent worth if left intact, has no metric in the economic calculations that support the industrialization of the wild. No company has to compensate the public for what it does to the commons, the forest, mountains, and rivers, which by definition belong to everyone. As long as there is a promise of revenue flows and employment, it merely requires permission to proceed. We take this as a given for it is the foundation of our system, the way commerce extracts value and profit in a resource-driven economy. But, if you think about it, especially from the perspective of so many other cultures, touched and inspired by quite different visions of life and land, it appears to be very odd and highly anomalous human behaviour.

Nearly all of what Davis implies in the statement above is relevant to the contemporary, and historic, situation in Guatemala. Just as it was in the early to mid 1900s with the United Fruit Company, where decisions made from boardrooms in Boston skyscrapers and the U.S. Congress (see Schlesinger and Kinzer 1983; Galeano 1988, 149–154) resulted in the complete reorganization—and disruption—of land and society in remote locations throughout the country, we see similar neo-colonial situations today, where space and livelihoods are controlled by transnational corporations based in Vancouver, Canada, and elsewhere (Fulmer et al. 2008), or, once again, from the insulated offices of the World Bank and IDB

(ADISMI 2007; Rights Action 2006; Holt-Gimenez 2007). And despite the humanistic, poverty-alleviating objectives that development agents may have for places like Guatemala (is it worth guessing the objectives of the corporations?), the strikingly different visions of 'progress,' between those few individuals who plan human-displacing projects such as massive infrastructure, mines, and energy development, and those who dwell upon the affected lands, will continue to generate increasingly violent conflict, and deplete cultural and natural heritage.

7.1 Concluding Thoughts

Despite the seemingly dismal scenario that I have scripted at times during the development of this research and book, my overall feeling after such in-depth analysis and fieldwork is positive and inspired. Rather than simply read about the dreadful state of marginalized peoples and the environment, which too often leads to negativity and disillusion, the firsthand glimpse I had into their struggles and history empowered me beyond any activity in which I have participated. I accredit this to the individuals I met, not just in Río Negro and Pacux, but also those in San Marcos resisting the Canadian-owned Marlin Mine, those in Quetzaltenango who lived through the state terror of the 1980s, and countless others who shared with me their stories and unrelenting passion for change. Nowhere had I heard such commitment for recognition and equality—though I don't doubt that this type of strength and dedication does exist in every corner of the world, even within the proximity of suburban Southern California, where such struggle appears non-existent. Even in Pacux, where the memories of such horrible, unthinkable acts linger, and the chances of attaining any form of reparation or prospects for an easier life seems at times unattainable, a noticeable force keeps survivors and their families moving forward, laughing, sharing, and fighting.

To conclude, I would like to reflect ever so briefly on an interaction I had early on in my ethnographic research, with an economist and professor conducting his own investigations in Central America. After disclosing to him my objectives and project design, the professor pleaded with me to consider new methods. "Simply talking to these people," he told me, "with no hard data… no one will take this study seriously!" He continued to try and persuade me to recreate my study based on comparative analysis, and the use of complex equations in order to prove that the people of Río Negro had been adversely affected.

Among certain circles (presumably, economists, and politicians who continue to design mega-projects as a means of development), he is probably correct in that my qualitative conclusions will hold little weight. I complete this study with no graphs, no quantitative measures to show changes in livelihood. My findings simply illuminate the 'perceptions' of those affected, and my own subjective observations and interpretations of the impacts that resulted from the dam and the violence. At the time, having not yet spoken to survivors and experiencing their lives in resettlement, my arguments against his proposed methods were weak, and I left the interaction feeling somewhat confused about how I would add strength, or 'validity,' to my research.

It was months later during one late afternoon sitting on the deck of the Centro Histórico, after so many days of talking with survivors, hearing stories of the past, present, and dreams to come, and watching life pass from my shaded perch above the reservoir, that the realization became apparent, quite overwhelmingly, and I remember scribbling in my notebook, after so much introspection: *There is no means by which to quantify what was lost here, in Río Negro, in terms of knowledge, culture, sacred ties, aesthetic beauty, ritual, and a connection unrecognized in economic theories. And even if the dam were not a complete failure, both economically and environmentally, would it have been worth it?*

It would be hypocritical for me to advocate, particularly as a North American never living for more than a few weeks without electricity, or produce grown in faraway tropical locals, or some form of petrol or minerals, that development as we know it must cease at all costs. And furthermore, it would be naïve, and irresponsible, to promote for a static cultural entity as some romantically believe the Maya to be. However, as cliché as it may sound, the activities and patterns of consumption that we assume as normal, and needed in our modern existence, must evolve. By illuminating the experiences of the Maya-Achí of Río Negro, and giving reference to the millions of others who succumb to similar fates, I aim to provide ample evidence for this claim, adding to the already existing call for holistic, multicultural, visions of growth, and 'development.'

7.2 Future Research

By no means does this research document everything there is to know about the Río Negro massacres, the Chixoy Dam, and the future of the survivors in regards to their struggle for reparations and justice. My work represents that of a dozen or so projects undertaken by outsiders interested in the story, and concerned with the human rights of those affected by Chixoy—in Río Negro as well as the other communities negatively impacted by the project. Continued interdisciplinary work with communities will be needed as long as reparations are unattained. As previously mentioned, despite the national and international awareness brought to this case by activists and academics, future projects involving involuntary displacement loom on the horizon for many communities in Guatemala. With the impressive number of social and natural scientists active in Guatemala and throughout Latin America, I situate myself within the camp of activist-oriented academics in proclaiming that it is our duty to alert the international community of what is happening in the far corners in which we study. Ideally, this research will inspire others to pursue the critical work needed to expose the effects of such developments, thus furthering the movement for justice and human rights.

References

ADISMI (Association for the Integral Development of San Miguel and Communities in Resistance in San Miguel Ixtahuacan) (2007). The negative impacts of the Goldcorp/Marlin mine in Maya Mam and Sipakapense Territory.

Alfred, T. (2009). *Colonialism and state dependency*. Project Communities in Crisis, Ottawa, Ontario: Prepared for the National Aboriginal Health Organization.

Bodley, J. H. (1990). *Victims of progress* (3rd ed.). Los Angeles, California: Mayfield Publishing Company.

Carmack, R. M. (Ed.). (1988). *Harvest of violence: The Maya Indians and the Guatemala crisis*. Norman, Oklahoma: University of Oklahoma Press.

Commission for Historical Clarification (CEH) (1999). *Guatemala: Memory of silence*. Report of the Commission for Historical Clarification Conclusions and Recommendations Guatemala. http://hrdata.aaas.org/ceh/report/english/

Davis, W. (2009). *The Wayfinders: Why ancient wisdom matters in the modern world*. Toronto, Ontario: Anansi Press.

Escobar, A. (1995). *Encountering development: The making and unmaking of the Third World*. Princeton, NJ: Princeton University Press.

Escobar, A. (2003). Displacement, development, and modernity in the Colombian Pacific. *International Social Science Journal, 55*(175), 157–167.

Fischer, E. (1996). The pan-Maya movement in global and local context. *Ph.D. Dissertation*. New Orleans, LA: Department of anthropology, Tulane University.

Fulmer, A. M., Godoy, Snodgrass, & Angelina & Neff, P. (2008). Indigenous rights, resistance, and the law: Lessons from a Guatemalan mine. *Latin American Politics and Society, 50*(4), 91–121.

Galeano, E. (1973). *Open veins of Latin America*. New York, NY: Monthly Review Press.

Galeano, E. (1988). *Memory of fire III. Century of the wind*. New York, NY: Norton and Company.

Grandin, G. (2004). *The last colonial massacre: Latin America in the Cold War*. Chicago, IL: University of Chicago Press.

Gutierrez, E. (1994). Rural upheaval and the survival of the Maya. *NACLA (North American Congress on Latin America) Report of the Americas*, 28.

Hamilton, S., & Fischer, E. F. (2005). Maya farmers and export agriculture in Highland Guatemala: Implications for development and labor relations. *Latin American Perspectives, 32* (33), 33–58.

Harris, C. (2003). *Making native space: Colonialism, resistance, and reserves in British Columbia*. Vancouver, Canada: UBC Press.

Handy, J. (1984). *Gift of the devil*. Toronto, ON: Between The Lines.

Holt-Giménez, E (2007). *Land-gold-reform: The territorial restructuring of Guatemala's highlands* (Development Report No. 16). Oakland, California: Food First Institute for food and development policy.

Ichon, A. (1978). *La Cuenca Media Del Río Chixoy* (Guatemala). Cuadernos de estudio Guatemaltecos 3. Guatemala City, Guatemala.

Imai, S., Mehranvar, L., & Sander, J. (2007). Breaching Indigenous law: Canadian mining in Guatemala. *Indigenous Law Journal, 6*(1), 102–129.

Johnston, B. R. (2005a). Volume One: Chixoy Dam Legacy Issues Study, Executive Summary: Consequential Damages and Reparation: Recommendations For Remedy. Santa Cruz, California: Center for Political Ecology. http://www.centerforpoliticalecology.org/chixoy.html

Johnston, B. R. (2005b). *Volume Two: Chixoy Dam Legacy Issues Study, Chixoy Dam legacy issues document review: Chronology of relevant events and actions*. Santa Cruz, California, USA: Center for Political Ecology.

Jonas, S. (1991). *The battle for Guatemala: Rebels, death squads and U.S. power*. New York, NY: Westview Press.
LaPlante, J. P. (2008, December 5). BC's new gold rush: Investment is risky in unceded land. *The Dominion*, Vancouver, BC. http://www.dominionpaper.ca/articles/2252
Lovell, W. G. (1988). Surviving Conquest: The Maya of Guatemala in Historical Perspective. *Latin American Research Review, 23*(2), 25–57.
Lynch, B. D. (2007). The Chixoy Dam and the Achi Maya. Violence, ignorance, and the politics of blame. Einaudi Center Working Paper. Cornell University. http://www.economyandsociety.org/events/conf_ethicsglobalization.shtml
Manz, B. (2005). *Paradise in ashes: A Guatemalan journey of courage, terror, and hope*. Los Angeles, CA: University of California Press.
Manz, B. (2008). The continuum of violence in post-war Guatemala. In Alisse Waterston (Ed.), *The Anthropology of War: Views from the frontline*. Oxford UK: Berghahn Books.
May, R. A. (1999). "Surviving all the changes is your destiny": Violence and popular movements in Guatemala. *Latin American Perspectives, 26,* 68–91.
Montejo, V. (2005). *Maya intellectual renaissance: Identity, representation, and leadership*. Austin, TX: University of Texas Press.
Osorio, J. T. (2003). *The Río Negro massacres*. Toronto, ON: Jesús Tecú Osorio & Rights Action.
Power, M. (2003). *Rethinking development geographies*. London, UK: Routledge.
Rights Action (2006). Xalala Dam Project: Water privatization, "development" & repression! [Chixoy Dam Revisited!]. Rights Action- Action Alert. http://www.rightsaction.org/urgent_com/Xala_Dam_111206.html
Schlesinger, S., & Kinzer, S. (1983). *Bitter Fruit: The untold story of the American coup in Guatemala*. Garden City, NY: Anchor Books.x.
Tauli-Corpuz, V. (2008). The concept of Indigenous peoples' self-determined development or development with identity and culture: Challenges and trajectories. Tebtebba Foundation, Bagulo City, Philippines.http://www.tebtebba.org/
Union of British Columbia Indian Chiefs. (2005). *Stolen Lands, Broken Promises: Researching the Indian Land Question in British Columbia*. Canada: Union of British Columbia Indian Chiefs.
Warren, K. B. (1998). *Indigenous movements and their critics: Pan-Maya activism in Guatemala*. Princeton NJ: Princeton University Press.
Windsor, J. E., & McVey, J. A. (2005). Annihilation of both place and sense of place: The experience of the Cheslatta T'En Canadian First Nation within the context of large-scale environmental projects. *Geographical Journal, 171*(2), 146–165.
Witness For Peace. (1996). *A people dammed: The impact of the World Bank Chixoy Hydroelectric Project in Guatemala*. USA: A Witness For Peace Publication. Washington D.C.
World Bank (2005). Operation Manual 4. 10: Indigenous Peoples. World Bank, Washington DC. Accessed May 1, 2009 http://web.worldbank.org/WBSITE/EXTERNAL/PROJECTS/EXTPOLICIES/EXTOPMANUAL/0,,contentMDK:20553653 ~ menuPK:4564185 ~ pagePK:64709096 ~ piPK:64709108 ~ theSitePK:502184,00.html
Wright, R. (2004). *A short history of progress*. Toronto, ON: House of Anansi Press.
Young, N. (2008). Radical neoliberalism in British Columbia: Remaking rural geographies. *Canadian Journal of Sociology*, 33(1) 1–36. 155.

Epilogue

There Is No Means By Which To Quantify What Was Lost In Río Negro

By Grahame Russell

Any update on the life struggle of the Mayan Achi people of Rio Negro can only be understood in the context of what Nathan wrote: "There is no means by which to quantify what was lost in Río Negro." There is no silver lining to the story of the Chixoy dam project. One cannot understate the enormity of the Chixoy dam crimes (massacres, killings and death by disease and hunger; complete dispossession and destruction of home, property and land) suffered by the Rio Negro villagers, crimes that in many ways continue today.

Perhaps an update can best be understood in terms of their survival, and then work and struggle to tell the truth and seek justice.

Survival

Most survivors live today—some with new children and grandchildren—in the former military controlled "model village" of Pacux, or on the dry mountainside above where the original Rio Negro community now lies under close to 100 m of Chixoy dam flood basin silt buildup and water.

Their ongoing life conditions of poverty, landlessness, joblessness, trauma and varying degrees of hopelessness are complicated and toxic, and are a direct result of the Chixoy dam crimes.

Yet, survive they do, and re-build, and tell the truth and demand justice. One might say it is "miraculous," but it is not. It is a testament to the Rio Negro survivors' (specifically) and the Mayan peoples' (generally) courage, dignity, and spiritual strength rooted deeply in their ancestors, Mother Earth, gods, and ceremonies.

N. Einbinder, *Dams, Displacement and Development*, SpringerBriefs in Latin American Studies, DOI 10.1007/978-3-319-51511-3_1

Truth

Since the first Chixoy Dam/Rio Negro mass grave exhumation, carried out in 1993–94 by the EAFG (Guatemala Team of Forensic Anthropology), precursor to the FAFG (Guatemalan Foundation of Forensic Anthropology), at the place known as 'Pakoshom' on a mountain ridge above the original village, the survivors—spearheaded by people Nathan interviews and discusses—have been truth-telling for any and everyone to hear.

Sadly, not surprisingly, the mainstream media, politicians, and "development" experts in the countries and institutions responsible for the Chixoy dam crimes—Guatemala and the U.S. primarily, the World Bank (WB) and Inter-American Development Bank (IDB)—pay as little attention as they can to this genocidal "development" project that is a skeleton in all of their closets.

Since this first Chixoy dam linked exhumation, the survivors have been demanding and supporting other mass grave exhumations (that are the fundamental truth-telling process in Guatemala today), and they have been initiating legal and quasi-legal efforts to seek justice (summarized below) in Guatemala and internationally. Survivor eye-witnesses, like Jesus Tecu Osorio and Carlos Chen Osorio, have written testimonial books; documentary films have been made; articles and reports written; survivor activists have traveled the world giving testimonials in conferences, and protesting at WB and IDB meetings. This book is part of the truth-telling process.

Justice

Despite widespread truth-telling about the Chixoy dam crimes, little justice has been achieved against any institution (IDB, WB) or government (Guatemalan, U.S.) for their roles as the intellectual and material authors of this "development" project and the crimes linked to it.

National Courts

There is an exception to this last statement. Since 1994, the Rio Negro survivors have courageously (receiving death threats) and, finally, successfully pressured the corrupted and dysfunctional Guatemalan legal system to put on trial, find guilty, and send to jail some nine former Civil Defense Patrollers (PAC) and military commissioners, mainly from the neighboring Mayan Achi village of Xococ.

On one level, this is an important achievement. In Guatemala's corrupted, manipulated legal system (still today), war crimes trials are few, and convictions fewer. However, on most levels, the jailing of nine of the lowest ranking material authors of the Chixoy dam crimes is proof of the impunity with which most material authors and all intellectual authors of the crimes act.

Not one single military personnel in the chain of command, from foot soldiers to generals, who ordered and carried out the Chixoy dam crimes was captured, tried, and sentenced. Not one official or program officer from the WB and IDB was subjected to any investigation into the role of the two "development" banks in partnering with the U.S. backed military regimes of Guatemala (1975–83) in planning and carrying out all aspects of the project.

Inter-American Court of Human Rights

Because only low-ranking PAC members were found guilty of the Chixoy dam crimes, and no officer in the chain of command was even detained, let alone tried, the Rio Negro survivors filed a petition with the Inter-American Commission of Human Rights, arguing impunity and lack of justice.

After years of delays, manipulations and threats to the Rio Negro survivors leading this quasi-legal struggle, on October 20, 2012, the Inter-American Court of Human Rights found the Guatemalan government responsible for the Chixoy dam crimes and ordered the government to investigate the Chixoy dam crimes; prosecute the perpetrators; search for the disappeared; carry out exhumations and identify the victims; publicly acknowledge its responsibility; build basic infrastructure and services for Rio Negro survivors in Pacux; implement projects to rescue the culture of the Maya Achi people; provide medical and psychological treatment to the victims; and pay compensation to surviving families for material and non-material damages suffered.

While this ruling of the Inter-American Court is also an important achievement of partial justice, it is noteworthy that the Court did not individualize responsibility—the ruling had no impact on the material and intellectual authors of the Chixoy dam crimes. Furthermore, the Inter-American Commission declined to investigate the roles and responsibilities of the IDB and WB, reinforcing their impunity. Finally, as of the writing of this epilogue, the government of Guatemala has not complied, at all, with the sentence.

Chixoy Dam Reparations Campaign

While not a legal process, this Reparations Campaign is an extraordinary achievement. After years of exhumations and other truth-telling struggles, the Rio Negro survivors united with approximately 30 other Mayan villages harmed by the Chixoy dam crimes to demand comprehensive reparations. (A clarification: the Inter-American Court sentence deals with the Rio Negro massacres directly linked to the Chixoy dam project; the Reparations Campaign deals with other losses and destructions caused by the project.)

As with the Inter-American Court case, it was only after years of delays, manipulations, and threats to the survivors leading this struggle, that on November 8, 2014, then president and former army general Otto Perez Molina (now in jail on corruption charges) formally apologized on behalf of the government for the human rights violations and sufferings caused by the Chixoy dam project, and signed into law Decree #378-2014, "the Public Policy of Reparations for Communities Affected by the Construction of the Chixoy hydro-electric dam project."

As of the writing of this epilogue, the government of Guatemala has paid out some of the $154,000,000 determined to be the amount to be paid in family compensation and re-building projects for the communities harmed and destroyed in varying degrees by the project.Notably, the WB and IDB easily resisted all pressures to be included in the Reparations Campaign investigation into roles and responsibilities—global "development" bank impunity reinforced, once again!

The Enormity of Global Impunity

Thus, 33 years later, a measure of reparations is being paid. While a measure of justice was provided by the Inter-American Court holding the Guatemalan government responsible (keep in mind that the government has yet to comply!), no justice has been done for the roles and responsibilities of the WB and IDB that promoted, designed, and implemented the project … (it must not be forgotten that the WB and IDB profited from their investments in this project).

No justice has been done for the roles and responsibilities of the military and political leaders of successive U.S.-backed military regimes that oversaw implementation of the project from 1975 to 1983.

Pointing out the reality of impunity for the authors and profiteers of the Chixoy dam crimes is to highlight the enormity of this global human problem. Today, governments and global banks, corporations, and investors push ahead with projects across the planet, always in the name of "development," violently displacing populations and destroying habitats, violating a wide range of individual and collective rights, and ravaging Mother Earth. The Chixoy dam was one such project, perhaps more violent and destructive than most.

The Enormity of Courage, Dignity and Strength

Pointing out the reality of impunity takes nothing way from the courage and dignity, strength, and vision of the Rio Negro survivors. Since I first met Carlos Chen, Jesus Tecu Osorio, Dona Antonia, Cristobal Osorio, and other Rio Negro survivors in 1994, I have been in awe of their heart, their spiritual strength, their dignity, their commitment to re-building lives, families and community, and their courageous, relentless struggle truth, and justice.

Since 1994, Rights Action (the U.S. and Canadian based non-profit organization I work with) has been a proud and ongoing supporter of dozens of development, human rights, and environmental projects initiated by the Rio Negro survivors. I am committed to continuing to support their work and struggles for re-building, truth and justice, and to continue to demand that the WB and IDB be held fully to account for their role in the Chixoy dam crimes.

Appendix

*All Photos were taken by the author.

(A) Annual procession to Pocoxom, in honor of the 177 women and children massacred there on March 13, 1982. Photos taken 1 km above the village of Río Negro and Chixoy reservoir, March 13, 2009

N. Einbinder, *Dams, Displacement and Development*, SpringerBriefs in Latin American Studies, DOI 10.1007/978-3-319-51511-3_1

(B) Milpa farming above the reservoir, near the village of Chicruz, roughly 5 km south of Río Negro. Due to the flooding of the valley, much of the agriculturally productive land is now underwater, forcing villagers to grow food on steep, eroding slopes

(C) Ruins at Cauinal, roughly 10 km upstream from Río Negro. During the wet season, when the reservoir is at capacity, these structures, as well as many others in the vicinity, are submerged

(D) A woman weaves outside her home in the INDE-built resettlement village of Pacux

(E) A street sign in Pacux, named after a massacre victim from Pocoxom

(F) Industrial pollution consisting of medical waste, sludge, and other garbage accumulates upstream from Río Negro. The children seen in the boat are from a small settlement near the Cahuinal Ruins

(G) View of the Chixoy Reservoir Centro Histórico, Río Negro

MIX
Papier aus verantwortungsvollen Quellen
Paper from responsible sources
FSC® C105338

If you have any concerns about our products,
you can contact us on
ProductSafety@springernature.com

In case Publisher is established outside the EU,
the EU authorized representative is:
Springer Nature Customer Service Center GmbH
Europaplatz 3, 69115 Heidelberg, Germany

Printed by Libri Plureos GmbH
in Hamburg, Germany